The Field Artillery

The Field Artillery

HISTORY AND SOURCEBOOK

Boyd L. Dastrup

GREENWOOD PRESS
Westport, Connecticut • London

Library of Congress Cataloging-in-Publication Data

Dastrup, Boyd L.
The field artillery : history and sourcebook / Boyd L. Dastrup.
p. cm.
Includes bibliographical references and index.
ISBN 0–313–27264–6 (alk. paper)
1. Artillery, Field and mountain—History. I. Title.
UF400.D37 1994
358.1′2′09—dc20 93–1142

British Library Cataloguing in Publication Data is available.

Library of Congress Catalog Card Number: 93–1142
ISBN: 0–313–27264–6

First published in 1994

Greenwood Press, 88 Post Road West, Westport, CT 06881
An imprint of Greenwood Publishing Group, Inc.

Printed in the United States of America

The paper used in this book complies with the Permanent Paper Standard issued by the National Information Standards Organization (Z39.48–1984).

10 9 8 7 6 5 4 3 2 1

Contents

Preface

Over the years, field artillery has played a vital part on the battlefield. For example, in the Persian Gulf War of 1991, field artillery possessed an unprecedented ability to deliver tremendous amounts of accurate firepower rapidly, neutralizing and destroying Iraqi targets. The capability of furnishing such support, however, has not always existed.

Once Europeans had learned about the explosive properties of gunpowder and decided that it had military possibilities, they introduced gunpowder artillery to defend or attack a castle or fortification and to support a field army on the battlefield. Although early siege artillery was heavy, ponderous, and dangerous, it readily breached castle and fortification walls that had been designed to withstand projectiles thrown by trebuchets, ballistae, and catapults; and it had greater range and firepower than the defender's artillery. This situation permitted the attacker to lay siege with impunity and eventually led to a revolution in fortification design to restore the balance between the besieger and besieged, and this in turn forced reforms in siegecraft.

In the fourteenth century, as siege artillery and garrison artillery were fighting for supremacy, Europeans started employing artillery on the battlefield. Over the next six centuries, field artillery went through four distinct revolutions. The initial revolution was the first use of artillery on the battlefield, which occurred when the English under Edward III (q.v.) utilized crude artillery pieces during the Battle of Crécy in 1346. During the following three hundred years, field artillery became more widespread on European battlefields because of its firepower.

However, early field artillery had limited use in combat. Because they were heavy, pieces with enough firepower to damage the enemy could not be easily moved to stay abreast of the other combat arms' movements to provide close support. In light of this, armies generally sited these cannon in fixed positions to fire a bombardment at the beginning of a battle to destroy the opponent's field artillery, to soften up the enemy, and to repeal an attack. In contrast, maneuverable artillery pieces with the ability to follow the shifts in the battle lacked sufficient firepower to inflict serious damage because their projectiles were too light. In short, field artillery support on the battlefield was problematic and often lacking.

Because of improved metallurgy and casting techniques and the drive for effective close support from field artillery, King Gustavus Adolphus (q.v.) of Sweden introduced mobile field guns early in the seventeenth century. However, the king's light fieldpieces did not have enough power to do serious damage, even though gun crews could maneuver them with relative ease to keep them abreast of the infantry advance. Years passed before the Europeans developed powerful, mobile field guns. Such weapons, often called true field artillery, first appeared in the mid-eighteenth century as the result of the monumental work of Jean-Baptiste Gribeauval (q.v.) of France, and revolutionized field artillery for the second time. Gribeauval's field artillery had the ability to stay abreast of shifts in the battle and simultaneously furnish firepower, setting a standard that other armies soon followed.

During the second quarter of the nineteenth century, industrialization and the appearance of rifled small arms with ranges of 600 to 1,000 yards led to the third revolution—the development of rifled field guns and indirect fire. Unable to move within canister range of 400 to 500 yards because of rifled small arms, field artillerymen had to improve the range of their cannon or lose the dominance that they had during the Napoleonic Wars. Technological advances during the Industrial Revolution enabled Europeans to introduce rifled field artillery with ranges of 4,000 to 5,000 yards during the 1840s and 1850s. By the 1870s the ranges of the most advanced rifled steel field artillery were far greater than the human eye could see. Along with the need to protect their fieldpieces from small arms fire, the increased ranges forced armies to develop indirect fire to take advantage of their field guns. In the process, armies abandoned direct fire, which had been employed since the introduction of artillery.

Between 1870 and 1900 European armies experimented with various methods of indirect fire. Eventually, the aiming point, forward observer, and compass method became the favored technique. Because it required reliable communications systems and trained personnel, effective indirect fire that could furnish close support proved to be extremely difficult, as the experiences of Western armies during World War I revealed.

With the realization that the field artillery could not provide effective close support, especially on a mobile battlefield, armies searched for ways to improve indirect fire. During the years following World War I, motorization, improved command, control, and communications systems, nuclear warheads, precision-guided munitions, aviation, and high technology collectively formed the fourth revolution in the development of the field artillery. Together, these advances made effective close support with indirect fire a reality and gave the field artillery unparalleled lethality and responsiveness.

This reference book focuses on field artillery to the exclusion of garrison, coast, antitank, and antiaircraft artillery and is divided into three parts. The history that forms Part I concentrates on the major developments in the field artillery but does not go into great detail. It is an overview and an introduction for additional study in the history of the field artillery. Part II is composed of bibliographical essays. The essays divide the literature into general military history to give the reader a background; field artillery history; technology; and tactics, doctrine, and organization. To keep the bibliography manageable, I have excluded articles and books with marginal utility and included only those of greater value. In some cases, works that cover long periods of time or span the topics of history, technology, tactics, doctrine, and organization are cited more than once. Part III consists of brief biographical profiles of people whose contributions were of key importance. The choice of people also depended on the availability of information, which was sparse in some instances. (Cross-references to this section are indicated by the abbreviation *q.v.*)

Appendices include a short chronology and comparative tables. Author and subject indexes are designed to make this reference easily accessible to academicians, professionals, students, and general researchers with varied interests.

I would like to thank Robin Higham, Edgar F. Raines, Jr., and Larry M. Kaplan for patiently reading the manuscript and providing their helpful insights to clarify points and avoid mistakes.

I want to thank my wife, Karen, for supporting me in this project and suffering through the long evening hours as I worked at the computer.

For any errors in fact or omissions, I take full responsibility.

Glossary

Adjust: To correct the aim of fire.

Bag Charge: A propelling charge for a gun or howitzer which is contained in cloth bags rather than in a metal cartridge case.

Barrage: System of artillery fire in which guns are laid to deliver a line of bursting shells in front of the advancing infantry.

Breech: The rear part of a cannon.

Caliber: The internal diameter of the gun barrel.

Cannon: A generic family name for all tube artillery.

Carriage: The support for the cannon.

Carrier Shell: A shell which carries a payload to the target, such as leaflets, submunitions, smoke, gas, and so forth.

Chamber: The part of the bore of a cannon that holds the propelling charge, especially when the chamber is a different size from the rest of the bore.

Close Support: The action of the artillery against targets that are sufficiently close to the supported maneuver arm to require detailed integration into the plan of movement of the maneuver arm.

Defilade: To shield the gun from enemy fire or observation by use of natural or artificial obstacles.

Elevation: The angle between the horizontal plane and the gun barrel.

Enfilade Fire: Fire striking the flank of the target, generally from a defilade position.

Grapeshot: A cluster of small iron balls grouped around a wooden spindle and covered by a heavy cloth netting.

Gun: An artillery piece that throws a shell along a relatively flat trajectory at a high velocity.

Howitzer: An artillery piece that throws a shell along a curved trajectory.

Indirect Fire: Artillery fire in which the target cannot be seen from the gun.

Limber: A two-wheeled vehicle to which the carriage is attached for support when being transported.

Massed Fire: The method of fire in which the guns are laid so that the falling shells are grouped on one particular point for maximum effect.

Mortar: An artillery piece designed to be fired at a high angle so that the projectile plunges down sharply onto the target.

Obturation: The sealing of the rear end of the gun (breech) to prevent unwanted escape of gas from the propelling charge.

Range: The distance from the gun to the target.

Shell: A hollow artillery projectile.

Shot: A solid artillery projectile.

Shrapnel: An artillery shell filled with steel balls that is exploded by a powder charge in flight.

Traverse: To move a gun horizontally in order to point at the target.

Part I

History

1

Early Artillery Developments, 1350–1763

Although gunpowder first appeared in China for incendiary purposes and later for use in rockets in the thirteenth century, and was most likely introduced into Europe by the Arabs, the Europeans quickly adapted the explosive for military purposes. Realizing that gunpowder could hurl a heavy object more efficiently than existing engines of war—the ballista, the catapult, and the trebuchet—the Europeans developed gunpowder artillery and soon became the leader in cannon design, production, and employment. Commencing early in the fourteenth century, the Europeans employed artillery of various sizes as siege pieces to knock down castle walls, as garrison weapons to defend castles and fortifications, and as field artillery, the term used by contemporaries to designate artillery employed on the battlefield. Supporting the infantry and cavalry, however, proved to be difficult for early field artillery. Because of the arduous task of moving heavy cannon with sufficient firepower to do serious damage to the enemy, armies did not maneuver them on the battlefield but sited them in fixed positions. In comparison, light artillery pieces of the time lacked adequate firepower to do significant damage, even though they could be manhandled and stay abreast of changes on the battlefield. Years would pass before the Europeans could combine firepower and mobility in a single field artillery piece to create true field artillery.

THE FORMATIVE YEARS

During the thirteenth and fourteenth centuries, the Europeans encountered gunpowder for the first time, after their contact with the Middle East during the Crusades and with the Moors in Spain. After traveling through Spain, Roger Bacon (q.v.) returned to England in the mid-thirteenth century, where he reputedly described the components of gunpowder. Based probably upon an Arabic or Oriental formula, his mixture consisted of saltpeter, sulfur, and powdered charcoal that made a lightning-like flash when it was ignited by a flame. Even though he concealed its composition in cipher to protect its destructive power, gunpowder did not remain a secret for long.[1]

According to legend, Berthold Schwarz (q.v.), a monk in Freiburg, Germany, discovered the propelling power of gunpowder early in the fourteenth century while he was mixing the explosive's ingredients in his laboratory. Schwarz's mixture allegedly could raise heavy lids off pots and demolish masonry when it was poured into cracks and heated.[2]

Even though Bacon and the legendary Schwarz most likely never fully appreciated the value of gunpowder as a means to propel projectiles, many military commanders in Europe visualized its potential and set out to harness it in cannon. By the 1340s the French, English, Italians, and Germans were employing gunpowder artillery of various sizes. The Europeans' first venture with artillery involved manufacturing vase-shaped, cast-bronze artillery, called *pots de fer* by the French and *vasi* by the Italians. These small muzzle-loading weapons threw one-inch iron balls or stones or crossbolts, were relatively portable, and were ignited when a hot iron rod was touched to a train of powder in the vent that in turn lighted the powder in the chamber. Because these cannon lacked power, Europeans had to position them close to the castle gates to do any damage and still had to rely upon trebuchets, ballistae, and catapults to knock down castle and fortification walls. Moreover, the casting process required highly skilled craftsmen, was expensive, and produced flaws in the metal, such as air holes, while the heat and force generated by the explosion rapidly wore out the cannon tube.[3]

Motivated by the requirement for more potent, durable, and less expensive gunpowder artillery for siege warfare, Europeans replaced their vase-shaped cannon in the fourteenth century with larger cannon that they called bombards. Gun founders made the initial bombards out of wooden staves and held them together with iron hoops. Searching for even more reliable weapons for siege work, Europeans later supplanted wooden bombards with muzzle-loading and breech-loading wrought-iron bombards. Because the small size of the foundries of the times and the inability to generate high temperatures prohibited casting large metal objects, blacksmiths forged wrought-iron bombards by welding iron staves together and then shrinking iron hoops over the staves for added strength. These wrought-iron bombards threw stones that weighed up to 500 pounds about 1,400 yards. For example, Dulle Griete, built in 1382 by Flemish artisans, had a 25-inch bore, was 16 feet long, weighed 13 tons, and fired a 700-pound granite ball. Probably manufactured in Mons about the middle of the fifteenth century, Mons Meg was 13 feet long, had a 20-inch bore, weighed 5 tons, could toss a 450-pound projectile about 1,400 yards, and was used by Scottish kings between 1455 and 1513 to reduce the castles of rebellious nobles. Because of the monstrous sizes of some of these bombards, they were laid on rugged mounts of heavy wood and were anchored firmly to the ground by stakes. Despite the bombards' power and ability to destroy castle walls, their tendency to explode and the exhausting production techniques encouraged the Europeans to search for better methods of manufacturing artillery and to retain their engines of war.[4]

After discovering that the same procedure employed for casting church bells could be utilized for making cannon, Europeans introduced cast-bronze (an alloy of copper and tin) bombards in the fifteenth century. Using molds to form a hollow cylinder, bell makers, the only people in preindustrial societies with experience in casting metal objects, fashioned bronze bombards and then smashed the mold. Because of this, no two bronze bombards were alike. Although wrought iron was less expensive than bronze, forging wrought-iron bombards was more laborious than casting bronze ones and was also more costly. In view of this, cast-bronze bombards soon became the favorite siege weapon in Europe even though the wrought-iron ones continued to be made into the sixteenth century. Even so, cast-bronze bombards' gigantic sizes, often 30 inches in diameter and 15 to 20 feet in length, made them immobile.[5]

In 1453, Mohammed II of Turkey, for example, helped bring an end to the Byzantine Empire when he attacked Constantinople with sixty-eight giant bombards. The biggest, called Basilica, weighed 35 tons, had a 30-inch bore, required sixty oxen and two hundred men to move it, and threw a 1,600-pound projectile about 2,000 yards. Cast for the sultan by Urban, a Hungarian cannon founder, Basilica took two hours to load. It lacked a carriage or any type of support and was aimed on a mound of dirt that sloped upward toward the enemy. Basilica and other huge bombards battered the city walls of Constantinople for forty-seven days and finally created a breach to allow the city to be stormed by infantry. Despite their lack of mobility, Mohammed II's and others' clumsy wrought-iron and cast-bronze bombards shattered castle walls and revolutionized siege warfare.[6]

Temporary civilian gun crews served the bombards. In many instances, a civilian craftsman who had cast the bombard sold his services to the highest bidder, functioned as its master gunner, and had as many assistants as required. The cannon that they fired were treacherous, and the powder was unstable. If the powder was rammed too tightly, it could not get enough oxygen to support rapid burning and would simply smolder until someone removed the projectile and loosened the powder with a rammer staff. Often after loosening, the powder ignited and sent the rammer staff shooting through the air. Between shots the gun crew had to sponge out the inside of the cannon to dampen any remaining powder charge. If a spark was missed, the crew member who ladled in the next charge would often be hurt by a premature explosion. Such dangerous work led artillerists to adopt Saint Barbara (q.v.), who protected people against storms and explosions, as their patron saint.[7]

As they worked to improve their dangerous bombards, Europeans pursued the development of smaller artillery pieces. In the fifteenth century, cannon manufacturers learned that much smaller artillery could do the same damage as a bombard if it had a strong barrel and could fire iron cannonballs. In light of this, cannon founders started casting breech-loading and muzzle-loading bronze cannon that were 8 to 13 feet long, had bores of 1 to 6 inches in diameter, and weighed 300 to 8,000 pounds. Yet, the inability to seal the breech (obturation) to prevent gases from the powder explosion from leaking, resulting from the lack of machining ability and primitive casting methods, discouraged the widespread use of breech-loading artillery and led Europeans to concentrate on manufacturing cast-bronze muzzleloaders. The size of these cannon made them more mobile than bombards and, therefore more threatening and appealing to employ.[8]

Simultaneously, Europeans experimented with different forms of gunpowder. At first Europeans used fine gunpowder (called serpentine) that was undependable and poor in quality because it had a tendency to disintegrate into its components after setting for some time or being jostled during movement. In their quest for better gunpowder, Europeans began forming it into small grains or corns early in the fifteenth century. Corn gunpowder burned more rapidly than serpentine because the exposed surface of the separate corns could burn at the same time to produce a more powerful explosion than that of serpentine gunpowder. The combination of corn gunpowder and bronze cannon gave Europeans relatively powerful, mobile pieces in the fifteenth century that could accompany a marching army for employment as a siege or battlefield weapon.[9]

To make artillery truly mobile and effective, however, required a suitable carriage, better-designed cannon, and improved gunnery. Through the mid-fifteenth century armies mounted their cannon on sleds and their smaller bombards on heavy carriages that were not much more than beams with an axle that joined the beams and two crude solid wood wheels. In the 1450s the Swiss introduced a special carriage for transportation. The carriage had two oversized wagon wheels that were connected by a stout axle upon which the cannon was lashed, and had curved timbers extending from the axle, called trails, that facilitated moving the cannon and that dug into the ground when the cannon was fired to partially counteract the recoil. Equipped with this revolutionary carriage, armies could pull siege cannon cross-country more easily than before and attack virtually any castle or fortification. In the meantime, the Dutch first cast trunnions, or knobs, on the sides of the cannon tube, an innovation that ended the practice of lashing the tube to the axle. Trunnions fit into sockets on the carriage and permitted elevating or depressing the cannon tube to any desired angle. The invention of the quoin, a wedge-shaped block upon which the breech rested, provided limited elevation adjustment and further improved laying the cannon. Together, corn gunpowder, more mobile carriages, trunnions, and quoins dramatically improved artillery by the end of the fifteenth century.[10]

Although many countries played critical roles in the development of artillery during the fifteenth century, the French assumed the lead in their efforts to rid their country of the English during the closing years of the Hundred Years' War. Under the Bureau brothers, Jean and Gaspard, the French established a permanent artillery organization and created an artillery that was superior to any others in Europe and that became the cornerstone of the French army. Taking advantage of improved metallurgy and corned gunpowder, the Bureau brothers' cannon battered down English-held castles with ease in the 1430s and 1440s to contrast with the lengthy five- to six-month sieges of Henry V's campaigns in Normandy against Rouen in 1418 and Cherbourg in 1419.[11]

Toward the end of the fifteenth century, the French made another significant contribution when they adopted horse-drawn artillery and the carriages first developed by the Swiss. French cannon could stay abreast of the infantry on the march and were more mobile than the slow-moving, oxen-drawn artillery of contemporary armies. When he invaded Italy in 1494 to make good his claim to the throne of Naples, Charles VIII of France had the latest in artillery technology. Jacques de Grenouilhac, the grand master of ordnance for the French army, assembled an impressive artillery train of mobile and powerful horse-drawn cannon that were manned by trained gun crews. Awed by the mobility and firepower of French siege cannon, first Florence and then the pope yielded after only token resistance. On one occasion when a fortress on the borders of the kingdom of Naples tried to oppose the invaders, French gun crews took only eight hours to reduce its walls to rubble. As Charles VIII's invasion confirmed, the new cannon were more mobile, could move cross-country with relative ease, and made medieval castles obsolete because they could not withstand the battering from the new heavy, mobile siege cannon.[12]

Improved gunnery techniques complemented developments in cannon and carriage design that Charles VIII had employed so effectively in the 1490s. In 1537 Niccolò Tartaglia (q.v.), an Italian mathematician, wrote the first scientific treatise on gunnery (*Nova Scientia*), in which he discussed the principles of ballistics. During the next ten years, he invented the gunner's quadrant, an aiming device, and demonstrated that a cannon shot the greatest distance when it was elevated to an angle of forty-five degrees. He also first described the continually curving trajectory of a cannonball and pointed out that it never did travel on an absolutely level course. Later, in the 1590s, Luis Collado, a Spanish mathematician and engineer, computed firing tables based on the use of the gunner's quadrant and recorded many new firing techniques. Tartaglia's and Collado's efforts made gunnery more scientific than before.[13]

Early in the sixteenth century, in the meantime, Europeans had added the mortar, a short, thick-walled cannon, to assist siege cannon. Apparently invented by a German, the mortar was mounted on a heavy wooden base and had a high trajectory to permit lobbing explosive bombs into fortifications. Along with cast-bronze cannon, which eventually made the bombard obsolete, mortars offered a new element of flexibility to the artillery and composed an effective combination of powerful siege weapons.[14]

Understanding that bronze cannon upset the delicate balance between the besieger and besieged. Europeans responded by mounting artillery in their castles. Consisting of varying calibers and sizes, garrison or fortress artillery provided counterbattery fire and shot at anyone trying to assault the walls. Generally, Europeans positioned short-range light cannon on the high walls of their medieval castles, but these weapons had difficulties reaching long-range siege pieces. When heavier cannon were mounted in the castle, their recoil shook the wall, weakening it and making breaching easier. For this reason, Europeans generally avoided arming their castles with heavy artillery. This, in turn, allowed the attacker to site siege cannon in the open with little concern for counterbattery fire to batter the defender with impunity.[15]

Influenced by the success of Charles VIII's cannon in the 1490s, Europeans remodeled their existing castles, but more importantly, commenced building fortresses with thick walls, emplacements for heavy cannon, and bastions and outworks to throw artillery cross fire upon the enemy and to be mutually supporting. These new fortifications appeared early in the sixteenth century in Italy, spread rapidly throughout the rest of Europe, and restored the balance between the besieger and besieged. By the 1520s fortifications based on the Italian model were capable of withstanding even the best-equipped attackers. However, their cost was enormous. Only the wealthiest states and cities could afford the cannon and the labor required to construct the fortifications.[16]

The emergence of heavily armed fortifications in the sixteenth and seventeenth centuries stimulated serious efforts to reform siegecraft. To provide cover from enemy weapons, the attackers constructed trenches with gabions (wicker baskets filled with earth), without any discernible system, from their main lines to the presumed weak point in the fortification wall. Because the attackers were unable to conceal their progress, the defenders could concentrate murderous fire on them and inflict large numbers of casualties. To protect the besieging force, Sébastien de Vauban (q.v.) of France refined siege warfare even more in the 1670s. Drawing upon the ideas of others, he built a series of parallels, connected them with zigzag trenches, and employed more artillery than ever before. The use of parallels protected the besiegers and their equipment, while the zigzag trenches made forward movement less hazardous. Because of its effectiveness, Vauban's system dominated siegecraft for the next 150 years.[17]

In the meantime, the technological advances that had made artillery successful in siege warfare also contributed to its accomplishments on the battlefield. Gunpowder artillery made its first appearance on the battlefield during the fourteenth century. At the Battle of Crécy in 1346, the English under Edward III (q.v.) deployed small wooden pieces that fired lead or iron balls and bolts. Because these cannon lacked carriages and were placed on the ground, they were immobile and unable to support the infantry and cavalry effectively as the battle shifted. The only thing that they did was to scare the horses with their explosions.[18]

Subsequent use of artillery on the battlefield did much more than just frighten the horses. During the Hussite Wars of Bohemia (1419–1424), John Ziska (q.v.) effectively employed small cannon mounted on carts in concert with infantry and cavalry to defeat opponents. With the development of carriage-mounted artillery later in the fifteenth century, however, Europeans started employing cannon more extensively on the battlefield than they had done during the previous century. During the Battle of Formigny in 1450, for example, the French placed artillery on the flanks of their infantry line and hit the charging English with enfilade fire. At the Battle of Castillon in 1453, the French poured murderous enfilade fire from bronze cannon onto the assaulting English. For most of the fifteenth century, European gun crews situated their cannon in static positions in front of the other arms and restricted their work to a preliminary bombardment, because their cannon were too heavy to be maneuvered easily when the battle became fluid. As a result, artillery still had a limited value on the battlefield even with the appearance of relatively mobile carriages in the fifteenth century. Even so, these early field cannon established their usefulness and laid the foundation for future employment of field artillery.[19]

In the pursuit of more potent artillery to support the infantry and cavalry, the French introduced lighter bronze cannon during the fifteenth century. The French mounted them on mobile carriages and developed the limber, a two-wheeled vehicle that was attached to the trails of the carriage for transport. During the Battle of Fornova in 1495, the French under Charles VIII not only moved their horse-drawn artillery with more celerity than the Italians could maneuver their heavy oxen-drawn cannon, but also had trained cannoneers who could rapidly load and traverse their weapons. In this battle and others early in the sixteenth century, the French effectively employed artillery against enemy personnel on the battlefield. Rather than assigning their artillery specific missions, however, the French persisted in using their cannon, regardless of their size, interchangeably in siege work or on the battlefield. Yet, those very French successes in battle with their cannon in the late fifteenth century and early sixteenth century led many Europeans to conclude that only lighter cannon were suitable for use on the battlefield.[20]

With French field artillery leading the way, European cannon started playing a greater role supporting the field army. Although European artillery still lacked mobility early in the sixteenth century because of weight and had a low rate of fire of about eight rounds an hour, Europeans still valued cannon on the battlefield. Because of this, they continued searching for more mobile artillery pieces. Interestingly, each manufacturer designed his own patterns for cannon and equipment with little concern for standardization. This gave European armies a motley collection of cannon and equipment that made resupply and repairs in the field difficult, because each piece required its own size of shot and because parts from one cannon could not be used to fix another piece.[21]

The great diversity of types and calibers of artillery created chaos. Realizing that the multiplicity of calibers hampered efficient operations and was also costly to maintain, the Hapsburg emperor Maxmilian I (q.v.) pursued significant reforms in the 1490s. To improve the efficiency of his artillery, he classified it as siege or field artillery. While heavier cannon made up his siege artillery, lighter ones formed his field artillery. In addition, the emperor ordered the general use of iron shot, promoted gunnery training, organized his artillery into a separate branch of the army, established foundries, and ensured that cannonballs were of uniform size and that all guns of the same type had the same caliber.[22]

Following Maxmilian's example, most European armies restructured their artillery during the sixteenth century. They reduced the number of calibers, standardized ammunition, and eliminated breechloaders. For example, in the 1540s the Portuguese restricted their artillery to seven calibers after experiencing the inconvenience of having too many. During the following decade, Henry II of France limited his artillery to six types of ordnance. By the end of the 1500s, European artillery was generally composed of cannon that threw 12- to 80-pound projectiles, culverins that threw 10- to 60-pound shot, bastard culverins that threw 2- to 8-pound shot, and mortars that lobbed 40- to 200-pound projectiles. For the most part, heavier pieces dominated siege warfare, while lighter ones were becoming more prevalent on the battlefield. Ammunition, in the meantime, was composed of iron shot, canister (a can filled with small projectiles), grapeshot (a cluster of iron balls held together by a cloth net), and explosive bomb (an iron shell filled with gunpowder). Equally important, by the end of the century the Europeans abandoned any further serious work on light breech-loading cannon because of the difficulties of finding a suitable method for sealing the breech.[23]

Lacking money for bronze artillery, in the meantime, Henry VIII of England turned to cast-iron cannon. In the mid-sixteenth century he imported French metallurgist Peter Baude and sent him to the Sussex iron foundry, where cast-iron cannonballs were being produced. Using a technique described by the Italians, Baude manufactured some of the first cast-iron cannon. Although cast-iron artillery was cheaper, bronze pieces remained the favorite of Europeans because they were lighter than those of similar caliber of cast-iron and because they were more durable than brittle cast-iron cannon and less likely to burst when fired. Notwithstanding these liabilities, the inexpensive cost of cast-iron cannon encouraged kings and princes to purchase them from the English or to make their own.[24]

Even though armies did not take to the field without artillery in the sixteenth century, light cannon of the time did not have sufficient striking power to do much damage, while heavier ones were still too cumbersome to be maneuvered easily with the other combat arms. In summary, cannon were restricted from effectively supporting the infantry and cavalry because of the inability to combine firepower and mobility in the same weapon.[25]

Attitudes during the last half of the sixteenth century also hindered the productive use of artillery on the battlefield. Oriented toward siege warfare rather than a war of movement where expensive armies could be destroyed easily, European armies had more heavy than light cannon in their inventories and positioned various sizes of cannon in front of the infantry line to fire a few preliminary salvos at the beginning of a battle. When the action developed, the artillery remained silent and stationary because European artillerymen lacked suitable weapons, tactics, and organization to facilitate maneuvering their guns.[26]

Notwithstanding these limitations, artillery had experienced much progress since it had been introduced. Huge bombards had replaced the small, vase-shaped cannon, while cast-bronze cannons had supplanted bombards. Because of advances in metallurgy, gunnery, powder making, ammunition, and cannon and carriage design, cannon of the late sixteenth century were more mobile and powerful than their ancestors and were making their presence felt more than ever before in siege warfare and field operations.

FOUNDATIONS OF TRUE FIELD ARTILLERY

Understanding the importance of combining mobility and firepower in the same artillery piece to support the other combat arms, and seeking to infuse more firepower into his army, King Gustavus Adolphus (q.v.) of Sweden laid the foundation for the appearance of true field artillery in his efforts to protect Protestantism and Swedish national security. In the 1620s he developed a leather-bound gun that threw a 3-pound projectile. The gun consisted of a copper tube screwed into an iron-bound brass breech. The copper tube was wrapped with alternate layers of cord and a plaster-like substance and was finally covered with varnished leather. Because the gun was so light (weighing about 650 pounds), two men could pull and fire it during battle, as the Polish campaigns of 1628–1629 demonstrated. The leather-bound gun's lack of power in comparison with metal cannon and its short life span inspired Gustavus to develop a bronze cannon of the same size and weight. In addition to being mobile, the king's bronze cannon could fire three rounds in the time that it took a musketeer to fire one. Even though Gustavus failed to produce a light, mobile artillery piece with firepower, his efforts represented the first substantial step in the quest for such a cannon.[27]

Gustavus's other military reforms dramatically changed his artillery. During the late 1620s, he replaced the traditional system of hiring civilian gunners by employing soldiers that were subject to military discipline, organized a permanent tactical artillery regiment of six companies, and assigned his artillery a specific function according to size and weight. Gustavus designated 24-pounders as siege artillery, 12-pounders as field artillery, and 4-pounders as regimental guns. In doing so, the king decided that only light, mobile guns would support the field army in combat and rationalized his artillery at the same time. With the help of Lennart Torstensson (q.v.), the Swedish artillery chief, Gustavus later added a 9-pounder to his field artillery. Equally important, the king formed his 9- and 12-pounders (called guns of position or field artillery) into batteries of five to ten guns on elevations behind the infantry line to mass fire, attached two 4-pounders to each regiment for close support, and employed grapeshot, case shot or canister, or cartridge ammunition (a bag or case holding the complete powder charge and in some cases containing the projectile) as antipersonnel rounds to smash hostile formations. The cartridge eliminated the need to ladle powder into the chamber between shots, provided consistently measured charges, and increased the rate of fire and accuracy.[28] Gustavus's innovations of the 1620s formed the basis for splitting his army's mobile artillery into two echelons of combat support (regimental guns and field artillery) and were eventually adopted by other Europeans armies years later.[29]

During the Thirty Years' War (1618–1648), Gustavus not only campaigned outside of Sweden (when he intervened in the wars being fought over religious and secular differences in the German-speaking states in Central Europe) but also maneuvered his regimental guns and massed artillery fire to produce devastating results and reinforce the significance of his reforms. At the Battle of Breitenfeld in Saxony in September 1631, Johann Tserclaes Reichsgraf (Imperial Count) von Tilly, the general of the Catholic League, dragged along heavy cannon that each required fourteen horses to move. After an opening artillery duel that was won by Gustavus's numerically superior artillery, having higher rates of fire with cartridge ammunition, the battle became mobile. Whereas Tilly's ponderous artillery remained in fixed positions throughout the battle, were captured by the Swedish army, and were even employed by the Swedes to fire on Tilly, Gustavus moved his regimental pieces around at will, massed fire from all of his artillery to tear gaping holes in Tilly's lines, and employed his artillery effectively with the infantry and cavalry on the battlefield to produce about twenty thousand enemy casualties. By doing this, he established a significant precedent for other Europeans to imitate.[30]

Following Gustavus's example, many Europeans improved their artillery during the remaining years of the seventeenth century. Besides starting to militarize their gun crews, the Europeans abandoned the practice of designating pieces by a name and began identifying them by size and weight to minimize confusion. For example, a cannon that threw a 4-pound ball ceased to be called a minion or falcon and became a 4-pounder. Yet, Europeans still did not assign their cannon specific functions based upon size and weight, such as field, garrison, or siege artillery, and therefore failed to adopt one of Gustavus's most important reforms. For the most part, bound by tradition, conservatism, and arsenals of cannon that would have been expensive to replace, generals continued dragging an ill-assorted collection of artillery of various sizes with them into battle, which limited their ability to adopt aggressive tactics and organize their artillery effectively for combat as Gustavus had done.[31]

Besides Gustavus's measures, perhaps the most far-reaching reforms in the seventeenth century were those introduced by Louis XIV of France. In 1671 he organized a regiment specifically for artillery duty and founded the first artillery school in Europe in 1690 to develop trained gun crews and to eliminate some of the mystery behind artillery. Louis's active interest in artillery and his numerous wars also encouraged the French to invent the elevating screw to simplify adjusting the barrel's elevation without moving the carriage. Another improvement, the prolonge, a system of ropes that was attached to the carriage, allowed gun crews to pull their pieces over difficult ground more easily than before. Finally, Louis's cannon had a priming tube filled with powder inserted into the touchhole to ignite the charge in the chamber and make firing safer.[32]

Louis's wars stimulated the inventiveness of his opponents as well. In their search for a versatile cannon, the Dutch developed the howitzer in the 1690s to throw larger projectiles than cannon could fire. The howitzer was a light, short-barrel artillery piece with a length between that of the mortar and that of the field gun. Because it employed a light powder charge in comparison to that used in a field cannon, it had a slower muzzle velocity than its bigger cousin and had a curved trajectory intermediate between cannon's flat trajectory and mortar's high-angle trajectory; it thereby gave the artillery three distinct types of pieces categorized by the trajectory of the projectile. Although howitzers were initially employed in siege work to lob exploding shells over walls, the Europeans soon adopted them for field operations to throw exploding projectiles into troop formations or hit soldiers behind temporary field fortifications.[33]

Inspired by the frequent wars of the eighteenth century, the drive to upgrade artillery continued unabated. After being appointed by the king of France to bring order to French artillery, General Jean-Florent de Vallière (q.v.) reduced the number of patterns and calibers to eliminate confusion and to stay abreast of foreign developments. In 1732 he standardized French cannon calibers as 4-, 8-, 12-, 16-, and 24-pounders, and added mortars. He also specified lengths, proportions, and weights for each and even decreed the methods of manufacture. All cannon of the same caliber were to be alike to simplify supply and repair and promote standardization. Interestingly, Vallière designed his artillery to fight from fixed positions and did not clearly designate them as field, siege, or garrison artillery. Even so, he gave the French the only unified artillery system in Europe.[34]

Confronting superior Prussian artillery fire during the War of the Austrian Succession (1740–1748), the Hapsburgs acted to improve their own. In the midst of the war, Prince Joseph Wenzel Liechtenstein (q.v.), director general in the Hapsburg army, established an artillery school for practical and theoretical instruction, militarized artillery personnel, organized three artillery brigades of eight companies each, outlined approved patterns for field artillery of 3-, 6-, and 12-pounders that were lighter and more mobile than their predecessors, and supplemented them with light howitzers. Liechtenstein also devised a rational system of deploying field artillery. He assigned the bulk of his 3-pounders in batteries of four to six guns to each infantry regiment as battalion guns, centralized the remaining 3-, 6-, and 12-pounders as guns of position under the chief of field artillery, who stood to the right of the army commander, and established solid shot and canister as the basic rounds, with the latter being the most effective against massed infantry, especially on the defense. Liechtenstein's reforms helped the Austrians gain battlefield parity in regard to artillery by the beginning of the Seven Years' War (1756–1763). In fact, the Prussians adopted the Austrian 12-pounder field gun, while the French were copying the Austrian light howitzers as late as 1803.[35]

In the meantime, under the direction of Field Marshal Count Peter Shuvalov, the seventh master general of the ordnance, the Russians modernized their artillery along the lines of their European counterparts. Otherwise, the Russians would have fallen behind their rivals in Europe.[36]

Despite advances in equipment, organization, and tactics during the early part of the eighteenth century, armies had to site their field artillery carefully prior to battle for effective support, because field cannon were still too heavy to be moved easily even in the set piece battles of the time. If positioning was done with wisdom, field guns, usually 6- to 12-pounders, could play a role in battle, while 3- and 4-pounder battalion guns were still issued to the infantry and manhandled during the advance. Even though these light guns supplied close support, the low weight of their solid shot limited their impact.[37]

Moreover, smoothbore field howitzers and cannon of the day were direct fire weapons. To hit the target the gunner had to see it. The gunner laid the piece by looking along the barrel and aligning it on the target. Traversing was done by moving carriage trails until the fieldpiece was lined on the target. Because of these characteristics, commanders usually placed their field artillery on the flanks of the supported troops or on elevations where the gunner could see the enemy, and placed their battalion guns between the infantry battalions for close support. Otherwise, gunners could not see their targets and could not fire, which meant that the infantry and cavalry would lack artillery support and had to attack on their own.[38]

New manufacturing techniques of the mid-eighteenth century paved the way for further improvements in artillery. In 1740 Jean Maritz, a Swiss engineer and gun manufacturer who had entered the employ of the French in 1734, revolutionized cannon production. He instituted the casting of cannon as a solid piece of metal, with the bore being drilled out, rather than pouring molten metal into a form with a core to create a hollow cylinder. Frequently, cannon cast by the latter method had irregularities in the alignment of the bore and circumference of the barrel because the core often shifted during the process of molding. Bores fashioned by Maritz's method made artillery pieces safer because the uniformity of the gun metal provided the same strength and thickness on every side of the explosion. This, complemented by more precisely cast round shot, reduced the windage (the space between the projectile and the bore) and the amount of gas that could escape around the projectile when the cannon was fired. Because of decreased windage, a small powder charge could propel a projectile with greater accuracy and velocity than a large charge in a cannon produced by the previous method. This advance simultaneously meant that cannon did not require thick tubes and could be lighter to enhance mobility.[39]

Although Maritz developed a boring machine, his successor and son, also named Jean Maritz, perfected the necessary machinery by the 1750s. As inspector general of gun foundries and forges for the French army, the younger Maritz installed cannon-boring machinery in all the royal arsenals in France. Soon other European states had adopted the new technique even though the French tried to keep the new method of manufacturing cannon a secret.[40]

In the meantime, Benjamin Robins (q.v.), an English mathematician and military engineer, examined exterior, interior, and terminal ballistics. He discovered many errors in the ballistic theories of Galileo and Isaac Newton outlined in the seventeenth century, abolished much of the myth and guesswork about the nature of gunpowder and the laws governing the flight of a projectile, and established the modern science of ballistics in the process. On the basis of his work, he published *New Principles of Gunnery* (1742), in which he explained the advantage of elongated projectiles, the value of rifling, and the impact of air resistance upon the flight of the cannonball. Later, he even recommended adopting rifled cannon and firearms. Although hunters employed rifled muskets, Europeans rejected equipping their armies with rifled weapons because they were difficult to load and not suitable for combat. Years passed before Europeans adopted rifled weapons for their armies, because they had to await the development of lathes and machining techniques.[41]

Building upon the technological advances of the early eighteenth century and trying to stay abreast of cannon development in other countries, Frederick the Great (q.v.) of Prussia restructured his artillery in the 1740s and 1750s. He categorized his cannon as siege or field artillery and formed artillery regiments for training, administration, and movement onto the battlefield. There, however, Frederick dispersed his lighter cannon along the infantry line as battalion guns. This practice left the artillery regiment with only the heavier guns that were massed in batteries in the center of the formation. At Mollwitz in 1741, for example, Frederick positioned his heavier pieces in front of the infantry center, distributed his lighter artillery along his line, and had difficulties employing his cannon effectively against the Austrians. Dissatisfied with his artillery support at Mollwitz and impressed with the Austrian artillery reforms of the 1740s, Frederick increased the number of 3- and 12-pounders in his army to almost six fieldpieces for every thousand men, an unprecedentedly high ratio, grouped his field artillery into batteries at critical points along the infantry line to mass fire at decisive points, and utilized howitzers to hit reverse slopes and shell enemies behind fortifications.[42]

Like Gustavus, Frederick appreciated speedy movement. Motivated by this concern, he worked to introduce more mobile artillery. He thought that foot artillery (the only kind of field artillery at the time), in which the gun crew walked beside the draft animals that pulled the cannon, was too slow and inadequate. In 1759 Frederick introduced horse artillery, in which the artillerists were soldiers and rode on the horses that towed the cannon to permit moving the weapon rapidly as the battle shifted and to serve as a strategic reserve or to support the cavalry. With the emergence of horse artillery, field artillery divided into two branches—horse and foot. Usually, Frederick positioned his horse artillery ahead of the infantry and trained his gun crews to mass their fire to smash a hole in the enemy's lines to allow the infantry to pour through. About 500 yards in front of the enemy lines, Frederick's gun crews dismounted and moved forward on foot, pushing their guns ahead of them, firing rapidly, and using canister or grapeshot during the latter part of their attack. The gun crews pressed forward until they were so close that they did not have to elevate their gun tubes to hit the enemy (point-blank range). Frederick's infantry then pushed through the cannon to attack the enemy.[43]

Frederick the Great revolutionized the field artillery, established the importance of massing fire even more than Gustavus Adolphus had done, and gave Prussia the lead in artillery development. As Frederick was introducing lighter guns, new organizations, and aggressive tactics to exploit his guns, his contemporaries generally kept their guns stationary during battle, dispersed them along the infantry line, and often did not have any cannoneers to serve the artillery. As a result, their artillery frequently sat idle during the heat of battle.[44]

Although tactical and organizational reforms, advances in technology, and better training for gun crews had improved the field artillery's ability to support the other combat arms by the mid–1700s, light, mobile pieces with firepower (true field artillery) still did not exist. The state of technology in the mid–1700s precluded combining firepower and mobility into one weapon. In view of this, European commanders echeloned their artillery. They sited their field artillery (6-, 8-, and 12-pounders) on hills overlooking the battlefield where they could see the target, moved these pieces only as needed because they were too heavy for easy maneuvering, and attached light battalion guns to the infantry for close support. Even though an artillery duel opened the battle to knock out the opponent's artillery, allowing the friendly infantry to attack unopposed by enemy artillery fire, and to soften up the enemy infantry, most artillerymen opposed extensive counterbattery work. They believed that it was counterproductive on the defense because it was difficult to hit an opposing fieldpiece and put it out of action. Rather, they desired to conserve their ammunition for the main mission of repelling enemy infantry and cavalry. On the offensive cannoneers employed their field guns to break up enemy infantry and supply counterbattery fire as required. The cavalry was still important, but massive volleys of musket and artillery generally determined who won the battle.[45]

In the meantime, the Russians introduced a bold technique for fire direction. During the 1750s, they experimented with a primitive form of observed indirect fire to hit targets that the gun crew and commander could not see by employing an observer to adjust fire. Yet, existing technology prevented widespread use of this method of fire, and years passed before effective observed indirect fire became a reality.[46]

As developments of the mid-eighteenth century suggested, the field artillery now had the ability to play a greater role on the battlefield than it had done in the previous four centuries. Horse artillery provided unprecedented mobility to permit shifting some of the artillery around the battlefield with relative ease to serve as a strategic reserve or to support the cavalry, while massing fire gave the field artillery the capacity to create greater damage than it had done previously. Yet, the field artillery had reached a plateau. Before true field artillery could appear, it was necessary to introduce lighter guns and equipment for even more mobility without sacrificing firepower in the process, while at the same time tactics and organization had to be developed to facilitate massing fire.

NOTES

1. Frank E. Comparato, *Age of Great Guns: Cannon Kings and Cannoneers Who Forged the Firepower of Artillery* (Harrisburg, PA: The Stackpole Company, 1965), 4; Oliver L. Spaulding, Hoffman Nickerson, and John W. Wright, *Warfare: A Study of Military Methods from the Earliest Times* (New York: Harcourt, Brace, and Company, 1925), 406; Richard A. Preston and Sydney F. Wise, *Men in Arms: A History of Warfare and Its Interrelationship with Western Society,* 4th ed. (New York: Holt, Rinehart, and Winston, 1979), 96.

2. Fairfax Downey, *Cannonade: Great Artillery Actions of History, The Famous Cannons, and the Master Gunners* (Garden City, NY: Doubleday and Company, Inc., 1966), 288; Comparato, *Age of Great Guns,* 3–4; Phillip E. Stevens, *Artillery through the Ages* (New York: Franklin Watts, Inc., 1966), 11–12; Albert Manucy, *Artillery through the Ages: A Short Illustrated History of Cannon, Emphasizing Types Used in America* (Washington, DC: Government Printing Office, 1949), 32.

3. Henry W.L. Hime, *The Origins of Artillery* (London: Longmans, Green, and Company, 1915), 123; Joseph Jobe, ed., *Guns: An Illustrated History of Artillery* (Greenwich, CT: New York Graphic Society, 1971), 11–12; Sir Charles Oman, *A History of the Art of War in the Middle Ages,* Vol. II, 2nd. ed. (New York: Burt Franklin, 1924), 224.

4. Carlo M. Cipolla, *Guns, Sails, and Empires: Technological Innovation and the Early Phases of European Expansion, 1400–1700* (New York: Pantheon Books, 1965, reprinted by Sunflower University Press, Manhattan, KS, 1985), 21–24; Theodore Wertime, *The Coming of the Age of Steel* (Chicago: University of Chicago Press, 1962), 64, 67; Manucy, *Artillery through the Ages,* 3–4; H.C.B. Rogers, *A History of Artillery* (Secaucus, NJ: The Citadel Press, 1975), 21–22.

5. Cipolla, *Guns, Sails, and Empires,* 22–23; Wertime, *The Coming of the Age of Steel,* 224–25; William H. McNeill, *The Pursuit of Power: Technology, Armed Force, and Society since* A.D. *1000* (Chicago: University of Chicago Press, 1982), 83–89; Bernard and Fawn Brodie, *From Crossbow to H-Bomb* (Bloomington, IN: Indiana University Press, 1973), 44–51; Stevens, *Artillery through the Ages,* 11–17; O.F.G Hogg, *Artillery: Its Origin, Heyday, and Decline* (Hamden, CT: Archon Books, 1970), 34–36, 212–13; Philippe Contamine, *War in the Middle Ages,* trans. by Michael Jones (London: Basil Blackwell, 1984), 142–46; Lynn Montross, *War through the Ages,* rev. ed. (New York: Harper and Row, 1960), 178–80.

6. Cipolla, *Guns, Sails, and Empires,* 22–23; Wertime, *The Coming of the Age of Steel,* 224–25; McNeill, *The Pursuit of Power,* 83–89; Brodie, *From Crossbow to H-Bomb,* 44–51; Stevens, *Artillery through the Ages,* 11–17; Hogg, *Artillery,* 34–36, 212–13; Montross, *War through the Ages,* 178–80; Contamine, *War in the Middle Ages,* 142–46; Downey, *Cannonade,* 23–35.

7. Manucy, *Artillery through the Ages,* 4–5; Stevens, *Artillery through the Ages,* 16–17.

8. H. W. Koch, *Medieval Warfare* (New York: Cresent Books, 1983), 206; Hogg,

Artillery, 35–36; McNeill, *The Pursuit of Power,* 89; Wertime, *The Coming of the Age of Steel,* 66; Manucy, *Artillery through the Ages,* 35.

9. Cipolla, *Guns, Sails, and Empires,* 27–28; Montross, *War through the Ages,* 180; McNeill, *The Pursuit of Power,* 87–89; Stevens, *Artillery through the Ages,* 11–17; Brodie, *From Crossbow to H-Bomb,* 44–53; Contamine, *War in the Middle Ages,* 145; Hogg, *Artillery,* 93.

10. Jobe, *Guns,* 18–19; F. L. Taylor, *The Art of War in Italy, 1494–1529* (Cambridge: Cambridge University Press, 1921), 83–86; J.F.C. Fuller, *Armament and History: A Study of the Influence of Armament on History from the Dawn of Classical Warfare to the Second World War* (New York: Charles Scribner's Sons, 1945), 83; Montross, *War through the Ages,* 186, 207; Hogg, *Artillery,* 98–99; Rogers, *A History of Artillery,* 24; McNeill, *The Pursuit of Power,* 87–89; O.F.G. Hogg, *English Artillery: 1326–1716* (London: Royal Artillery Institution, 1963), 10.

11. Russell F. Weigley, *The Age of Battles: The Quest for Decisive Warfare from Breitenfeld to Waterloo* (Bloomington, IN: Indiana University Press, 1991), 38–39.

12. McNeill, *The Pursuit of Power,* 88–89; Weigley, *Age of Battles,* 38–39; Taylor, *The Art of War in Italy,* 81–82, 88–90, 94; Spaulding, Nickerson, and Wright, *Warfare,* 416–17.

13. Comparato, *Age of Great Gunners,* 9.

14. Fuller, *Armament and History,* 83; Montross, *War through the Ages,* 180; McNeill, *The Pursuit of Power,* 87–89; Stevens, *Artillery through the Ages,* 11–17; Brodie, *From Crossbow to H-Bomb,* 44–53; Contamine, *War in the Middle Ages,* 145; Hogg, *Artillery,* 94.

15. Comparato, *Age of Great Guns,* 7–8; McNeill, *The Pursuit of Power,* 89–95; Contamine, *War in the Middle Ages,* 201–7.

16. Comparato, *Age of Great Guns,* 7–8; McNeill, *The Pursuit of Power,* 91–95; Contamine, *War in the Middle Ages,* 201–7.

17. McNeill, *The Pursuit of Power,* 128; Christopher Duffy, *Siege Warfare: The Fortress in the Early Modern World, 1494–1660* (London: Routledge and Kegan Paul, 1979), 22, 52, 54, 71, 84, 93–95, 98; Henry Guerlac, "Vauban: The Impact of Science on War," in Edward Mead Earle, ed., *Makers of Modern Strategy: Military Thought from Machiavelli to Hitler* (Princeton, NJ: Princeton University Press, 1971), 39–40; Henry Guerlac, "Vauban: The Impact of Science on War," in Peter Paret, ed., *Makers of Modern Strategy: From Machiavelli to the Nuclear Age* (Princeton, NJ: Princeton University Press, 1986), 79–80; Ian V. Hogg, *Fortress: A History of Military Defense* (New York: St. Martin's Press, 1975), 51–52; Taylor, *The Art of War in Italy,* 95–96.

18. Jobe, *Guns,* 12; Rogers, *A History of Artillery,* 16.

19. Hime, *The Origins of Artillery,* 129–33; Taylor, *The Art of War in Italy,* 99; Downey, *Cannonade,* 1–12; Hogg, *English Artillery,* 196–200; Stevens, *Artillery through the Ages,* 14; Rogers, *A History of Artillery,* 43; Montross, *War through the Ages,* 180, 181, 186, 187, 190, 207; Contamine, *War in the Middle Ages,* 198–200; M.G.A. Vale, "New Techniques and Old Ideals: The Impact of Artillery on War and Chivalry at the End of the Hundred Years War," in C. T. Allmand, ed., *War, Literature, and Politics in the Late Middle Ages* (New York: Barnes and Noble Books, 1976), p. 67.

20. Taylor, *The Art of War in Italy,* 83–93; Cipolla, *Guns, Sails, and Empires,* 74; Montross, *War through the Ages,* 208–10, 273; Rogers, *A History of Artillery,* 24, 25–27.

21. Cipolla, *Guns, Sails, and Empires,* 28–29; Hogg, *Artillery,* 24–25; Hogg, *English*

Artillery, 40; Manucy, *Artillery through the Ages,* 6; McNeill, *The Pursuit of Power,* 166; Rogers, *A History of Artillery,* 35–46; Montross, *War through the Ages,* 238, 273.

22. Jobe, *Guns,* 26–31.

23. Manucy, *Artillery through the Ages,* 6; Stevens, *Artillery through the Ages,* 35–36; Jobe, *Guns,* 41–54.

24. Hogg, *Artillery,* 50–53; Stevens, *Artillery through the Ages,* 20; Brodie, *From Crossbow to H-Bomb,* 52; Rogers, *A History of Artillery,* 28; Hogg, *English Artillery,* 20.

25. Stevens, *Artillery through the Ages,* 35–36; Hogg, *Artillery,* 56–57; Cipolla, *Guns, Sails, and Empires,* 28–29; Jobe, *Guns,* 32, 38–44.

26. Hogg, *Artillery,* 24–25, 100–101; Hogg, *English Artillery,* 40; Manucy, *Artillery through the Ages,* 6; McNeill, *The Pursuit of Power,* 166; Rogers, *A History of Artillery,* 35–46; Montross, *War through the Ages,* 238, 273.

27. Cipolla, *Guns, Sails, and Empires,* 74–75; Stevens, *Artillery through the Ages,* 22; Rogers, *A History of Artillery,* 39–41; Jobe, *Guns,* 70; Weigley, *The Age of Battles,* 4.

28. Manucy, *Artillery through the Ages,* 7; Comparato, *Age of Great Guns,* 9; Stevens, *Artillery through the Ages,* 21–26; Rogers, *A History of Artillery,* 39–41; Jobe, *Guns,* 70.

29. B. P. Hughes, *Firepower: Weapons Effectiveness on the Battlefield, 1630–1850* (London: Arms and Armour Press, 1974), 14; Downey, *Cannonade,* 69–79.

30. Stevens, *Artillery through the Ages,* 23–26; Weigley, *The Age of Battles,* 14–23; McNeill, *The Pursuit of Power,* 119–23.

31. Brodie, *From Crossbow to H-Bomb,* 53; Rogers, *A History of Artillery,* 52; Manucy, *Artillery through the Ages,* 9, 32, 56; Harold L. Peterson, *Round Shot and Rammers* (Harrisburg, PA: Stackpole Books, 1969), p. 36.

32. Comparato, *Age of Great Guns,* 10–11; Brodie, *From Crossbow to H-Bomb,* 53; Rogers, *A History of Artillery,* 52; Manucy, *Artillery through the Ages,* 9, 32, 56; Peterson, *Round Shot and Rammers,* 36.

33. Hogg, *Artillery,* 94–95; Manucy, *Artillery through the Ages,* 9, 32, 56; Peterson, *Round Shot and Rammers,* 36.

34. McNeill, *The Pursuit of Power,* 166; Hogg, *Artillery,* 94–95; Manucy, *Artillery through the Ages,* 9, 32, 56; Peterson, *Round Shot and Rammers,* 36.

35. Gunther E. Rothenberg, *The Art of Warfare in the Age of Napoleon* (Bloomington, IN: Indiana University Press, 1978), 25; Peterson, *Round Shot and Rammers,* 33, 48–51; Weigley, *The Age of Battles,* 271.

36. Christopher Bellamy, *Red God of War: Soviet Artillery and Rocket Forces* (London: Brassey's Defence Publishers, 1986), p. 15.

37. McNeill, *The Pursuit of Power,* 167–70; Peterson, *Round Shot and Rammers,* 33, 48–51; Rothenberg, *The Art of Warfare in the Age of Napoleon,* 15, 25; Rogers, *A History of Artillery,* 33–36.

38. B. P. Hughes, *British Smoothbore Artillery: The Muzzle Loading Artillery of the 18th and 19th Centuries* (Harrisburg, PA: Stackpole Books, 1969), 20, 61.

39. McNeill, *The Pursuit of Power,* 167–70; Weigley, *The Age of Battles,* 271.

40. McNeill, *The Pursuit of Power,* 167.

41. Rogers, *A History of Artillery,* 59, 64; Brodie, *From Crossbow to H-Bomb,* 101, 103; Kenneth Brookes, *Battle Thunder: The Story of Britain's Artillery* (Reading, England: Osprey Publishing Ltd., 1973), 50.

42. Sir Edward S. May, *Achievements of Field Artillery* (Woolwich, England: Royal Artillery Institution, 1893), 4–17; Robert B. Asprey, *Frederick the Great: The Magnificent Enigma* (New York: Ticknor and Fields, 1986), 285; Stevens, *Artillery through the Ages,* 28–29, 51–52; Rothenberg, *The Art of Warfare in the Age of Napoleon,* 25; Weigley, *The Age of Battles,* 271.

43. Asprey, *Frederick the Great,* 509; Stevens, *Artillery through the Ages,* 27–33; Rogers, *A History of Artillery,* 53–57; Manucy, *Artillery through the Ages,* 10–11; Rothenberg, *The Art of Warfare in the Age of Napoleon,* 24–25.

44. Comparato, *Age of Great Guns,* 12; Rothenberg, *The Art of Warfare in the Age of Napoleon,* 25; Stevens, *Artillery through the Ages,* 27–33; Rogers, *A History of Artillery,* 53–57; Manucy, *Artillery through the Ages,* 10–11.

45. Rothenberg, *The Art of Warfare in the Age of Napoleon,* 14–15; William E. Birkhimer, *Historical Sketch of the Organization, Administration, Material and Tactics of the Artillery, United States Army* (James Chapman, 1884, reprint ed., Westport, CT: Greenwood Press, 1968), 10; Jac Weller, "The Artillery of the American Revolution," *Military Collector and Historian,* Fall 1956, 97; Philip J. Haythornthwaite, *Weapons and Equipment of the Napoleonic Wars* (Poole, England: Blandford Press, 1979), 64; Louis de Tousard, *American Artillerist's Companion* (1809, reprint ed., Westport, CT: Greenwood Press, 1968), 111–15.

46. Bellamy, *Red God of War,* 16.

2

From Gribeauval to Rifled Field Artillery, 1764–1865

Between 1764 and 1865, the field artillery underwent extensive changes that took it to new heights of performance. During the latter years of the eighteenth century, Europeans developed true field artillery by combining firepower and mobility in the same piece and created new forms of organization and new tactics to take advantage of the new technology. Within years after mobile, powerful smoothbore artillery had appeared, however, the introduction of rifled field artillery in the 1840s and 1850s threatened to make the former obsolete and transform tactics and organization even more.

INTRODUCTION OF TRUE FIELD ARTILLERY

Prodded by the need to make field artillery more effective on the battlefield, serious reformers of the last four decades of the eighteenth century introduced vast improvements. With a firm determination to keep French field artillery abreast of its Prussian and Austrian counterparts and to restore it to the position of predominance it had under Charles VIII and his successors during the sixteenth century, Jean-Baptiste Gribeauval (q.v.) of France, who had served several years in the Austrian army, established an artillery system that became the mainstay of Napoleon's army thirty years later. After returning to France from Austria in the 1760s, Gribeauval was commissioned to rationalize French artillery. Using Austrian artillery as a model, Gribeauval, who became inspector of artillery in 1764, designated his artillery as field, siege, garrison, and coast artillery and introduced standardized carriages, limbers, ammunition chests, and tools. For the field artillery he selected 6-inch howitzers and 4-, 8-, and 12-pounder cannon. He intended his 4-pounder to serve as a battalion gun and made his 8-pounder and 6-inch howitzer the backbone of French field artillery by concentrating them at the brigade, while the 12-pounder was for heavy support in the general reserve. By doing this, Gribeauval divided his field artillery into battalion, brigade, and army guns and built upon the organizational precedents of Gustavus Adolphus (q.v.) and Frederick the Great (q.v.).[1]

Material improvements made possible Gribeauval's organizational reforms. Concerned with the mobility of his field artillery, the Frenchman took advantage of recent advances in metallurgy and gun founding. He reduced the weight of the gun barrel by approximately 50 percent of that of older French field guns of equivalent power, but retained the firepower. By combining mobility and firepower into one cannon, Gribeauval introduced true field artillery. In addition, he harnessed the horses in pairs rather than in file and replaced hardwood axles with iron ones for more durability. He installed a calibrated rear sight and graduated tangent sight at the breech that could be set to compensate for the dropping trajectory of the projectile. To make aiming easier and to make use of the sights, he included a elevating screw beneath the breech to facilitate raising and lowering the muzzle. He also supplied his cannon with canister, grapeshot, and explosive shell and adopted cartridge ammunition that combined the projectile and powder charge into one unit for better accuracy and rate of fire. Equally important, the Frenchman reduced the windage by 50 percent below earlier French artillery. The decrease in windage lengthened the range of the field gun because the explosive gases could not escape as easily around the projectile upon firing, and it reduced wear at the same time because the projectile bounced down the bore to a lesser degree than before. Even though Gribeauval's reforms greatly enhanced the field artillery, fierce opposition from conservative French officers delayed the French army from approving his system until Gribeauval had become first inspector general of artillery in the French army in 1776.[2]

In addition to his technical reforms, Gribeauval revamped the gun crew. He replaced civilian drivers with soldiers, who were subject to military discipline, and emphasized training. To attain the necessary precision, a gun crew practiced its movements until they had become automatic. Each man of a normal fourteen-man crew of the time had a number and performed a specific task rapidly at an exact time. The first six crew members, called matrosses, worked the drag rope, also called a prolonge, to maneuver the gun around the battlefield. The other eight, who were more skilled, sponged, loaded, rammed, aimed, and fired their piece. If they were highly trained, they could fire two to three aimed shots a minute. Since casualties during battle would deplete the number of men, each man on the gun had to know the duties of the others. Crews also drilled arduously with each other to work harmoniously as a team to smother enemy batteries with an overpowering volume of fire.[3]

Simultaneously, Gribeauval established schools to teach artillery officers the theoretical aspects of gunnery and tactics to strengthen their status as professionals. Besides establishing Gribeauval as a leader in artillery innovations, the Frenchman's reforms made French field artillery the most advanced in Europe and the model for others to imitate, even though Gribeauval's improvements were not fully felt until Napoleon's time.[4]

Chevalier Jean Du Teil (q.v.) of France quickly grasped the significance of the introduction of lighter materiel. Realizing that the Gribeauval system had dramatically improved mobility, Du Teil wrote *On the Use of the New Artillery* (1778), proposing a more meaningful role for French field artillery in battle. He suggested that tactics similar to those of siege warfare could be employed on the battlefield. Fieldpieces could be grouped in large numbers to breach the enemy line with their fire, allowing the infantry to advance. By exploiting Gribeauval's mobile pieces, the French could open the battle with massive field artillery barrages to shatter the opponent and then maneuver their field guns as the battle shifted to mass them at critical points and times. Equally important, Du Teil advocated employing field artillery against troop formations and not artillery. The difficulty of putting an artillery piece out of action made artillery duels simply unprofitable. Consequently, they should be avoided if possible.[5]

Du Teil's ideas slowly gained adherents because changes in tactics also meant creating new artillery organizations and abandoning tradition. At the opening of the Napoleonic Wars, the French army still divided its field artillery into battalion guns and guns of position as had been the practice in European armies for years. Battalion guns supplied direct support to the infantry, generally battalions or regiments, and were manhandled, often by untrained infantrymen, to keep them abreast of the infantry advance. Impressed with Du Teil's thinking, many French artillerymen advised the army to discard its battalion guns (generally 4-pounders) and championed forming all of the army's field artillery into batteries to mass fire rather than dispersing pieces along the infantry line, diluting firepower and encouraging the piecemealing of guns into action. Convinced of the need for better artillery support by centralizing command and control, the French abandoned their battalion guns and attached their lighter fieldpieces to the division and the heavier ones to the corps by the beginning of the nineteenth century. On the defense French field artillery repelled infantry and cavalry charges, while on the offense it softened up enemy infantry formations. As the offense unfolded, gun crews moved their cannon to enfilade the enemy or to find suitable targets. Although Du Teil and other French artillerymen promoted the creation of huge batteries to blast the enemy line, forming such batteries did not become the norm until after 1800, when Napoleon and other European generals began grouping their field guns into large batteries to crush the enemy with imposing displays of firepower.[6]

With the advent of powerful, mobile field guns, French commanders embraced aggressive tactics for their field artillery on the offense. Alexandre Antoine Hureau de Senarmont, one of Napoleon's corps artillery commanders, advocated the bold use of field artillery and great concentrations of guns in large batteries as promoted by Du Teil. At the Battle of Friedland in 1807, Senarmont assertively pushed his field artillery ahead of the friendly infantry line to within canister range (about 500 yards) to soften up the Russian line with withering fire and established a tactical breakthrough with field artillery in the process. As the French right was collapsing in the face of a determined Russian counterattack, Senarmont split his guns into a heavy reserve and two main units. He advanced his guns to about 500 yards, unlimbered, and fired, and repeated this process until his cannon were about 100 yards from the enemy's infantry line. By this time, French field artillery was dominating the battle and firing with impunity because Russian infantry were blocking (masking) the fields of fire for their supporting artillery. Unhindered by enemy counterbattery work, Senarmont continued moving his guns forward until they were within 60 yards of the Russians. From this position he hit the enemy line with furious canister fire to help Napoleon defeat the Russians. The relatively short effective ranges of smoothbore muskets (50 to 100 yards) permitted such determined tactics by Senarmont because smoothbore field guns of the time had greater effective ranges, even with canister, than muskets, making cannon an aggressive offensive weapon.[7]

To keep up with their continental rivals, the British had to modernize their field artillery. In 1759 John Muller (q.v.) published *A Treatise on Artillery,* which eventually inspired the British to develop lighter tubes and carriages for greater mobility after overcoming determined opposition from the master general and Board of Ordnance.[8] The work of General Sir William Congreve (the father of the developer of rockets by the same name) with artillery carriages further enhanced British field artillery. In 1792 he introduced the block-trail carriage. He replaced the two brackets that formed the trails of the carriage with a solid beam of wood to reduce the weight and to permit cutting sharper corners, and simultaneously shifted the center of gravity forward. This enabled one person to traverse the trail more easily than before with the aid of a handspike inserted into the end of the trail. At the same time, Congreve built a limber with an ammunition chest, and an ammunition wagon, with both seating cannoneers. Mounted artillery, as Congreve's system was soon called, enhanced the mobility of British field batteries because crew members rode on the limber and ammunition wagon into battle and arrived fresh and ready to serve their weapons and because the batteries could follow the shifts in battle more easily. Equally important, Congreve's limber, carriage, and ammunition wagon gave the field artillery three branches (foot, horse, and mounted) and made Gribeauval's system of field (foot) artillery in which the gun crew walked beside the cannon obsolete because it was less mobile.[9]

Even so, years passed before most European and American armies even recognized the superiority of the British system over Gribeauval's. The successes of the French system during the Napoleonic Wars blinded most armies to its weakness and discouraged searching for anything better. In fact, the French did not supplant the Gribeauval system with one comparable to the British style of carriage, limber, and ammunition wagon until 1827, after they were finally convinced that their own was inferior. The Americans later adopted a system similar to the British one in the 1830s and 1840s, but this came only after heated debates within the U.S. Army and after the British system had proved itself to be superior.[10]

The appearance of lighter field artillery materiel paved the way for the development of new organizations and tactics during the latter decades of the eighteenth century. Along with the lighter equipment, the procedure of employing the prolonge to join the limber and carriage together and using the limber horses to drag the fieldpiece around the battlefield allowed the Europeans to reduce the number of personnel in the gun crew. They abandoned the untrained matrosses, retained only seven to eight men in a gun crew, and drilled them arduously to perform their functions with a high degree of skill. Generally, number one stood at the right of the muzzle with the sponge and rammer to push home the powder charge and round and to sponge out the bore after a shot. Number two positioned himself on the opposite side of the muzzle and placed the powder charge and ammunition in the bore. Number three stood at the right of the breech and thumbed the vent, while number one sponged and rammed. This prevented smoldering pieces of cartridge bags from being forced into the vent and helped to create a suction when the sponge was withdrawn to aid in extinguishing any sparks that might have survived the sponging and to prevent a premature explosion. Number three also primed the cannon after the charge and round had been rammed home by sticking a priming tube filled with a mixture of gunpowder, saltpeter, and sulfur in the vent or by pouring loose powder into the vent. Number four held the lighted portfire, a thin cylindrical case of paper that contained a quick-burning composition of gunpowder, saltpeter, and sulfur and that had been lighted by touching it to the linstock. The linstock held the slow match that was three or more loosely woven strands of hemp boiled in lees of old wine, or a solution of wood ash or saltpeter, and was bound together with an outer layer of hemp. Upon command, number four fired the piece by touching the portfire to the primer in the vent. Number five manned the handspike that fitted into a socket in the trial transom to point the gun in the direction ordered by the officer in charge. The rest of the crew helped bring ammunition from the ammunition chest to the gun.[11]

Organizational reforms went beyond the battery formation and drill. Like the French, the British perceived that they obtained greater efficiency and flexibility by maintaining their field artillery under centralized control and abandoned the battalion gun concept, even though they continued to attach batteries to battalions on occasion to support the infantry and cavalry. When the Russians introduced powerful, mobile fieldpieces in 1806, they abolished the custom of assigning guns to the regiment. The Russians gave the division an artillery brigade and the corps an artillery general to command all of the corps's artillery and, like most European armies of the early nineteenth century, stressed massing fire and aggressively employing their field artillery as an offensive weapon *par excellence*.[12]

To facilitate massing fire with their mobile, bronze smoothbore field guns,

European commanders had to organize artillery reserves. Because of the limited effective ranges of the fieldpieces of the time (about 500 to 1,000 yards), commanders had to move their guns around the battlefield to mass fire and could not simply shift the direction of fire to mass fire on a particular target. To minimize the need for moving the guns along the infantry line, commanders held back one fourth to one third of them in the form of an artillery reserve. They sited these cannon in batteries in strategic positions behind the infantry to support an advance or repel a charge as needed, which meant that not all of an army's firepower was available at the beginning of a battle because some was being deliberately held back to mass fire at critical times and places.[13]

The search for mobile and powerful field artillery concurrently motivated European armies to adopt horse artillery as Frederick the Great had done in the 1750s. In 1793 the British formed their first horse artillery units and armed them with 6-pounders, 12-pounders, and 5.5-inch howitzers. The following year, the French established horse artillery as an independent elite corps. Each battery had six to eight pieces and consisted of four officers and seventy-two men. In the meantime, the Swedes and Russians organized their own horse artillery. The Europeans developed horse artillery for rapid movement, especially to support the cavalry, and generally equipped it with lighter weapons than foot or mounted artillery to ensure mobility. As combat action in the 1790s and after indicated, horse artillery effectively bolstered an attack or covered a retreat. Yet, its expense encouraged most European armies to retain foot artillery, which was less costly, as the major portion of their field artillery. Even so, field artillery with its three branches that complemented each other—foot, mounted, and horse—gave European armies unparalleled firepower and mobility.[14]

The development during the latter years of the eighteenth century and the early years of the nineteenth century of light materiel and field cannon that combined firepower and mobility stimulated armies to centralize command and control of their field artillery more than ever before, and simultaneously promoted massing fire in an unprecedented fashion. This combination of improved organization, tactics, and technology revolutionized the field artillery. As the Napoleonic Wars attested, field artillery played a greater role on the battlefield than it had done before because it could be maneuvered around the battlefield with relative ease to mass fire as needed and clear the way for the infantry to advance.[15]

Even the Americans, who were fearful of being dragged into the wars raging in Europe during the 1790s to stop revolutionary France from dominating the continent, took action to improve their artillery. Because of the XYZ Affair in 1798, which almost led to war between France and the United States, the American secretary of war, James McHenry, encouraged the United States to develop a system of artillery comparable to its European counterparts. When the crisis with France passed, Congress quickly abandoned McHenry's recommendations. Prior to leaving office in 1800, however, McHenry urged adopting horse artillery and training artillerymen in the latest tactics and drills, but the government's frugality and the fear of a trained army prevented the United States from seriously considering his advice. Nevertheless, his successor, Henry Dearborn, fielded new carriages based upon Gribeauval patterns and cast-iron cannon and appointed Captain George Peters in 1808 to command the U.S. Army's first horse artillery battery. Unfortunately, the Americans disbanded the battery the following year because of the prohibitive cost of maintaining the unit.[16]

Although equipment, tactics, and organization were slowly changing at the beginning of the nineteenth century, cannon technology had made mixed progress. Fieldpieces were lighter and more mobile and made out of better metal than their ancestors, but they were still black powder, inaccurate, bronze smoothbore weapons with effective ranges of 500 to 1,000 yards, were still direct fire weapons, and utilized canister, shell, shot, and occasionally grapeshot as their standard loads. Yet, fixed ammunition, which contained the projectile and powder charge in a single unit, was becoming more popular because it increased the rate of fire and accuracy. For the most part, European gun crews employed fixed ammunition when rapid firing was required and relied upon semifixed ammunition, in which the projectile and bag of powder were loaded separately, for most of their firing because it was less expensive.[17]

Searching for an antipersonnel projectile with a greater range than canister, Lieutenant Henry Shrapnel (q.v.) of Great Britain introduced spherical case in 1784 after several years of work. Unlike explosive shell, which burst into several fragments (five to seven), spherical case was filled with iron pellets and a small charge to detonate the round. When the fuze exploded the case, the contents continued traveling forward in a cone-shaped pattern. Yet, spherical case, which was later renamed shrapnel after its inventor, had one important drawback that discouraged employment. Because of friction heat created by the intermingling of balls and powder in flight, the projectile often blew up prematurely. Despite this, the British army adopted spherical case in 1804 after successful trials in 1803, promoted Shrapnel to lieutenant colonel, and employed it against the French in the Peninsular War. The round was an immediate success because it greatly enhanced the power of British field artillery by allowing the guns to hit enemy personnel with antipersonnel rounds at greater ranges than canister fire permitted.[18]

Works on rockets paralleled that on cannon artillery. Although rockets had existed for centuries, Europeans had forsaken them for cannon by the end of the fourteenth century because the latter were more accurate and reliable. Intense rocket bombardments by the Indian army against the British during the sieges of Seringapatam, India, in 1791–1792 and 1799, however, impressed the British so much that they searched for someone to manufacture rockets for their army in their drive to defeat the French. Because of the lack of any experts in Great Britain and because of his own interest in rockets, Colonel William Congreve (q.v.), the son of the developer of mounted artillery, improved on the Indian rocket. Using his own money, Congreve created the world's first rocket weapon system of 3- to 300-pounder rockets. He instituted mass production of rockets, improved their accuracy, and even devised rocket tactics. Besides developing incendiary rockets, Congreve manufactured ones with shrapnel warheads. During the siege of Boulogne in 1806, Congreve tested his rockets for the first time in battle. After setting the city ablaze, Congreve enthusiastically wrote that they would drastically change military tactics. Even so, Congreve's rockets were inaccurate and suitable only for use against unfortified targets in area bombardment.[19]

Notwithstanding this, for a time the British army shared Congreve's enthusiasm. It produced a rocket that could carry an explosive powder charge more than two miles, with the rocket exploding on landing by means of a time fuze, and later employed rockets to bombard Copenhagen in 1807 and Fort McHenry in the War of 1812. Stimulated by the British successes with rockets having ranges that were about twice those of contemporary smoothbore cannon, other European armies organized rocket units. This interest in rockets led to the publication of books, a large number of articles, and numerous patents during the first half of the nineteenth century. Spectacular developments with rifled artillery in the 1860s, however, caused rockets, including those produced by William Hale (q.v.), which were superior to all others, to fall out of favor.[20]

Improved cannon and rocket technology, tactics, and organization that appeared late in the eighteenth century and early in the nineteenth century gave the field artillery a greater role in combat than it had ever had, as the Napoleonic Wars indicated. With the ability to be maneuvered around the battlefield with relative ease and with firepower, the field artillery advanced to new heights and demonstrated its capacity to provide close support and to be an effective combat arm.

THE ECLIPSE OF SMOOTHBORE FIELD ARTILLERY

Between 1815 and 1865, Europeans participated in a prodigious industrial expansion that touched all aspects of life, including the field artillery. The steamship, the railroad, and the telegraph transformed communications and logistics, while improved ammunition, ballistics, and rifling greatly altered the field artillery.

Early in the nineteenth century, Europeans adopted the percussion cap to ignite a charge in a weapon. Following the development of fulminate of mercury in the 1790s, which could be detonated by concussion, a sports-loving Scottish clergyman, the Reverend Alexander Forsyth, finally perfected a percussion system of ignition in 1807. Forsyth employed a powder that would flash through the touchhole of a gun and ignite the powder chamber when it was struck by a hammer blow. When he invented the percussion cap for muskets and cannon in 1814, Joshua Shaw of Philadelphia took the next logical step. Initially he used iron, then pewter, and finally copper. Because converting flintlock muskets was easy, the percussion cap came into general use about 1820 for infantry weapons. The percussion cap also helped solve the chronic problem of misfire because of wind and wet weather.[21]

Years passed, however, before the percussion cap was adopted for field artillery use. Portfires and linstocks, rods entwined with a slow-burning rope called a slow match, were still carried for firing cannon as late as the 1850s and 1860s; however, gun crews employed them only when percussion-lock tubes, which required the blow of a hammer upon a percussion cap, and friction primers were not available, or when the new devices failed to work. The friction primer was composed of a copper tube with a serrated wire running through it, friction compound, and musket powder. Both the percussion tube and the friction primer were placed in the vent, and a lanyard, a long piece of cord, was attached to a loop at the end of the wire or attached to the percussion hammer. A quick pull on the lanyard dragged the serrated wire across the friction compound or tripped the hammer to fire the cannon.[22]

In the meantime, notable progress with ammunition and fuzes occurred. By the 1840s field artillerymen preferred fixed ammunition for shot, shell, canister, and spherical case to separate-loading ammunition because it increased the rate of fire and simplified loading the cannon. Even though most armies still had grapeshot in their inventories, they favored spherical case because it had a greater range than grapeshot. They also preferred canister to grapeshot because it was more effective at close ranges. Equally important, grapeshot damaged the bore of smoothbore bronze cannon and was slowly being discarded as a field artillery munition. In the meantime, Captain Charles Bormann of Belgium invented a fuze that contained a train of powder in the form of an arc that led to the powder charge in the shell. Bormann marked off the arc in quarter-second intervals to give it a total burning time of five seconds. By properly setting the Bormann fuze, gunners could detonate the shell more consistently over the heads of the enemy than they could with wooden fuzes. Yet, wooden fuzes were still used. A wooden fuze was tapered with a hole with portfire composition in it. The outside of the fuze was marked with lines to indicate the burning time. To set the fuze, gunners cut it at the desired line for the time duration and then inserted it into the projectile.[23]

Although better primers, improved fuzes, and the general acceptance of fixed ammunition enhanced the effectiveness of smoothbore bronze field guns during the years following the Napoleonic Wars, rifled weapons threatened to eclipse smoothbore artillery. In light of smoothbore artillery's performance on Napoleonic battlefields, many Europeans found it to be indispensable. From Benjamin Robins's (q.v.) work of the 1740s, gun manufacturers of the early 1800s knew that a spinning, elongated projectile from a rifled weapon was more accurate than a free-falling spherical shot from a smoothbore fieldpiece. Yet, foundries and machine shops of the early 1800s were unable to do the precision work that was critical for producing rifled cannon. If a round was just a few thousandths of an inch too thick or if the groove had the slightest irregularity, a projectile might jam and burst the tube, and this discouraged employing rifled artillery.[24]

Improved technology produced by industrialization and the appearance of rifled small arms in the 1830s and 1840s stimulated European inventors to adapt rifling to field artillery. Created by William Greener of Great Britain but improved upon by Captain Claude E. Minié of the French army, the expansive cylindro-conoidal bullet exploited rifling in a musket. The bullet had a hollow base so that when it was fired it would automatically expand to engage the rifling. The Minié bullet, as it was called, radically altered gun design and made rifling practical for the military for the first time because it permitted loading a rifled musket as rapidly as a smoothbore musket. Equally important, rifled muskets increased the infantry's range of fire to a range of 300 to 600 yards, in contrast to the previous effective range of 50 to 100 yards, with some even having ranges of 1,200 yards. Because small rifled arms had longer ranges than smoothbore muskets and greater rates of fire than smoothbore field artillery, muzzle-loading smoothbore cannon suddenly became antiquated. Such long ranges of rifled small arms, even though the latter were not totally accepted by European armies, meant that gun crews employing canister could not attack infantry formations with impunity, as they had done during the Napoleonic Wars, and that field artillery was no longer the dominant combat arm.[25]

To keep field artillery abreast of the developments in small arms, Europeans turned to rifled pieces. After centuries of neglect, Europeans started producing rifled muzzleloaders in the 1840s and 1850s. Because the projectile had to fit the bore snugly to engage the rifling so as to produce the spin, the initial rifled muzzleloaders were difficult to load and had low rates of fire. Realizing that breech loading was the only answer and that it reduced gun crew fatigue, Europeans experimented with various ways of designing a breech mechanism that would not be damaged by the gas pressure generated by the powder explosion and that would not leak gases. In 1846 Major Giovanni Cavalli of Italy developed the first workable rifled breechloader. He constructed a cast-iron gun with a spiral groove in the tube that ran from one end to the other and created an elongated projectile with two lugs to fit the grooves.[26] Nine years later, Sir Joseph Whitworth (q.v.) of Great Britain produced a rifled breechloader and muzzleloader, each with a hexagonal bore. Meanwhile, Sir William Armstrong (q.v.) of Great Britain manufactured a rifled breechloader. Taking advantage of existing technology, the Russians also experimented with rifled breechloaders in the 1850s, as did Baron Martin Wahrendorff of Sweden. The spinning action produced by the rifling gave rifled field cannon longer ranges and greater accuracy than smoothbores. By the early 1860s rifled field artillery had an average range of approximately 4,000 yards, whereas the average smoothbore had a range of 1,000 yards.[27]

The Italian war of 1859 demonstrated the superiority of rifled field artillery over smoothbores. With their greater ranges and incomparable accuracy rifled French field guns compelled the Austrians to position their cannon behind their infantry to protect them from hostile counterbattery fire. From these locations Austrian field guns could not hit the French nor support their own infantry. Encouraged by the effectiveness of their field cannon, French gun crews confidently bombarded the Austrians with little concern for hostile counterbattery work and helped the French to defeat the Austrians decisively.[28]

The search for better gun metal, meanwhile, also spurred significant improvements in field artillery. During the 1830s and 1840s, gun manufacturers changed from producing cast-iron guns to wrought-iron ones. Although wrought-iron cannon were better, several disastrous explosions with them in the 1840s discouraged employing them and prodded gun manufacturers to find an even better metal. The Krupps of Prussia exhibited a few rifled steel breechloaders at the Great Exhibition in London in 1851. The guns were light and had a tensile strength about four times that of cast iron and twice that of wrought iron. Four years later, the Krupps displayed even larger steel breechloaders at the Paris Exhibition. Impressed with the guns, the king of Prussia eventually ordered three hundred of them in 1858 and made his army the first to be equipped with rifled steel breechloaders. Most European and American field artillerymen, however, displayed less enthusiasm than the Prussians. Because several Krupp guns burst during trials, field artillerymen generally regarded steel as too brittle. Moreover, breechloaders leaked gases at the breech, were dangerous, and were more expensive than smoothbores. Aware of these problems, armies avoided equipping their field artillery with steel breechloaders and retained smoothbore bronze muzzleloaders and a few rifled wrought-iron muzzleloaders.[29]

Henry Bessemer (q.v.) of Great Britain made steel a reality for field artillery when he discovered a way to process it in the 1850s. By blowing air through the molten steel, he could regulate its chemical content more exactly and produce it more cheaply than before. The Bessemer process, as a result, made the large-scale production of steel possible and made existing methods of producing steel obsolete.[30]

Although the Bessemer process revolutionized steel production and made steel field artillery more of a reality, some inventors still clung to traditional metals. In the United States Captain Thomas J. Rodman (q.v.) of the U.S. Army cast iron cannon over a water-cooled core. Because the molten iron next to the core cooled faster than the succeeding layers, it exaggerated the shrinkage process and made the interior metal more dense than the outside. This caused his iron gun to be stronger than those cast as a solid block of iron or bronze and then bored out, or those cast around a sand core to create a hollow tube that was smoothed out by machining. Employing his technique, Rodman introduced a superior method of casting iron or bronze artillery and developed a whole family of Columbiad-type smoothbore guns for coastal defense, known as Rodman guns, that saw considerable action during the American Civil War.[31]

Another notable advancement about the same time came with the appearance of hooped or built-up field guns. In the 1820s A. Thiery of the French army conducted experiments with hooped guns by shrinking a wrought-iron envelope over a cast-iron barrel. In the 1840s, Daniel Threadwell of the United States manufactured a few hooped guns for the United States government by using the same technique. In the meantime, Sir William Armstrong utilized the built-up principle for his guns. Armstrong's pieces were built around a core by winding iron strips around the barrel lining, frequently made of steel, or by sweating hoops of iron around the core to build the gun in a series of layers. Sweating involved heating a hoop of metal to make it expand, and then slipping it over the already assembled parts of the gun. The hoop shrank as it cooled but not back to its room temperature dimensions. Rather, lasting internal tension squeezed the exterior bank tightly against the interior layers and created a force to oppose the explosion. This system of construction produced stronger guns than forging them out of a solid block of metal did.[32]

Captain Robert P. Parrott (q.v.) of the U.S. Army developed a different type of hooped rifled artillery. Working at the West Point Foundry in the state of New York, Parrott reinforced the chamber of a cast-iron cannon with a wrought-iron hoop. This method enabled a cast-iron gun to withstand the strong powder charges required to throw heavy elongated projectiles.[33]

Before rifled artillery could be truly effective, it required suitable rounds. The major problem confronting inventors involved designing a projectile that would contact the rifling so that it would spin. Yet, the round could not fit the bore so tightly as to make loading difficult. Some inventors developed rounds with studs to engage the rifling, while Whitworth manufactured a hexagonal projectile to be fired from a hexagonal bore. Such rounds were impractical and difficult to load in rifled muzzleloaders and were also expensive. Taking steps to make rifled ammunition cheaper, Captain Parrott produced a round with an expanding cast-iron cup built into the base. When the fieldpiece was fired, the gases pressed the cup against the rifling to spin the round. Later, in the early 1860s, a round with a copper or brass rotating band was developed that expanded to engage the rifling when the explosion gases hit it. Although many other kinds of ammunition were developed, most field artillerymen preferred expanding cup or rotating band ammunition and employed percussion and time fuzes to detonate their rounds.[34]

Interestingly, as many inventors were developing rifled muzzleloaders and breechloaders and suitable ammunition, Napoleon III (q.v.) of France designed a 12-pounder gun-howitzer that represented the high point of smoothbore artillery technology. Seeking to reduce the diversity of his artillery, Napoleon III adopted the 12-pounder fieldpiece during the mid–1850s because it combined the firepower of his heavier field guns and the mobility of his lighter ones. With the same goal that Napoleon had, the United States later introduced the 12-pounder Napoleon for service in 1857.[35]

The new field artillery met its first real test during the American Civil War of 1861–1865. Even though field artillery performed effectively during the war, it lost its role as an aggressive offensive weapon. Because of rifled muskets, gun crews had to position their batteries behind the infantry to keep them out of range of small arms fire. Also, crews had problems maneuvering their field guns over the difficult terrain of Civil War battlefields because the weapons were still relatively heavy (weighing up to 2,000 pounds). As a result, the field artillery often failed to furnish close support to the infantry, had problems massing fire on the offense, and lost the offensive role that it had during the Napoleonic Wars and the United States's war with Mexico in 1846–1848. With some exceptions, the rushing of field artillery to within canister range disappeared. In contrast, field artillery still played a significant defensive role because gun crews could concentrate their guns more easily to repel enemy charges. For example, at Malvern Hill in July 1862, the Union army massed its field artillery virtually hub-to-hub on rising ground within about a half mile of the Confederate army. Because the Confederates were unable to mass their artillery on the offense effectively, Union batteries fired with impunity and tore apart the Confederate advance.[36]

Interestingly, rifled muzzleloaders and breechloaders failed to supplant smoothbore muzzleloaders such as the 12-pounder Napoleon as the preferred weapon during the American Civil War for several reasons, even though they had greater ranges and better accuracy. Because wooded, difficult terrain in America restricted fields of fire, field artillerymen could not see the distance that rifled pieces could fire and had to fight at close ranges. Moreover, the extent of human eyesight (approximately 1,700 yards) and the visible horizon discouraged the employment of rifled pieces because the guns were still direct fire weapons, which meant that gun crews had to see the target to hit it. If the target was beyond the horizon or human eyesight or if something covered the line of sight, gun crews could not engage the target effectively. Hitting an unseen target on a mobile battlefield involved using observed indirect fire techniques that had not yet been perfected, even though the Russians had experimented with it in the mid-eighteenth century and Europeans had used it during their sieges in the Crimean War of the 1850s.[37]

More powerful and accurate rifled field guns appeared during the middle of the nineteenth century and threatened to supplant bronze smoothbores, but European and American cannoneers could not exploit the new technology effectively. Terrain and the limits of human eyesight thwarted the adoption of rifled artillery, while technological problems, especially with rifled breechloaders, discouraged their use. Even though technology had changed by the 1860s, Napoleonic field artillerymen would have still felt comfortable. After all, massing fire was still accomplished by using concentrations of direct fire guns. New techniques, tactics, and forms of organization were required before European and American armies could effectively employ rifled field guns on the battlefield. Until then, rifled field guns were simply impractical.

NOTES

1. Patrick Griffith, *French Artillery* (London: Almark Publishing Company Ltd., 1976), 7–8; James E. Hicks, *French Military Weapons: 1717–1938* (New Milford, CT: N. Flayderman and Company, 1964), 140–43; Joseph Jobe, ed., *Guns: An Illustrated History of Artillery* (Greenwich, CT: New York Graphic Society, 1971), 65; Russell F.

Weigley, *The Age of Battles: The Quest for Decisive Warfare from Breitenfeld to Waterloo* (Bloomington, IN: Indiana University Press, 1991), 272–73.

2. William H. McNeill, *The Pursuit of Power: Technology, Armed Force, and Society since* A.D. *1000* (Chicago: University of Chicago Press, 1982), 170–71; Harold L. Peterson, *Round Shot and Rammers* (Harrisburg, PA: Stackpole Books, 1969), 51–54, 74; Albert Manucy, *Artillery through the Ages: A Short Illustrated History of Cannon, Emphasizing Types Used in America* (Washington, DC: Government Printing Office, 1949), 11–12; Gunther E. Rothenberg, *The Art of Warfare in the Age of Napoleon* (Bloomington, IN: Indiana University Press, 1980), 25–28.

3. Peterson, *Round Shot and Rammers,* 64, 66; McNeill, *The Pursuit of Power,* 170–71.

4. McNeill, *The Pursuit of Power,* 170–71.

5. Robert S. Quimby, *The Background of Napoleonic Warfare* (New York: Columbia University Press, 1957), 88–89; Michael Howard, *War in European History* (New York: Oxford University Press, 1976), 78; Hew Strachan, *European Armies and the Conduct of War* (Boston: George Allen and Unwin, 1983), 33–34; John A. Lynn, *The Bayonets of the Republic: Motivation and Tactics of the Army of Revolutionary France, 1791–1794* (Urbana, IL: University of Illinois Press, 1984), 209–15; Rothenberg, *The Art of Warfare in the Age of Napoleon,* 23.

6. Rothenberg, *The Art of Warfare in the Age of Napoleon,* 74, 75, 107; Lynn, *The Bayonets of the Republic,* 209–11; Strachan, *European Armies and the Conduct of War,* 42–43; Quimby, *The Background of Napoleonic Warfare,* 296–97.

7. Griffith, *French Artillery,* 37–41; Weigley, *Age of Great Battles,* 406; Rothenberg, *The Art of Warfare in the Age of Napoleon,* 78–79.

8. John Muller, *A Treatise of Artillery* with preface by Harold L. Peterson, (reprinted by Museum Restoration Service, Ottawa, Ontario, 1965), preface; H.C.B. Rogers, *A History of Artillery* (Secaucus, NJ: The Citadel Press, 1975), 59–61, 63, 65, 68–71.

9. Rogers, *A History of Artillery,* 59–60, 83–84; B. P. Hughes, *British Smoothbore Artillery: The Muzzle Loading Artillery of the 18th and 19th Centuries* (Harrisburg, PA: Stackpole Books, 1969), 68–70; Hicks, *French Military Weapons,* 144–45; Boyd L. Dastrup, *King of Battle: A Branch History of the U.S. Army's Field Artillery* (Fort Monroe, VA: Office of the Command Historian, U.S. Army Training and Doctrine Command, 1992), 66.

10. Rogers, *A History of Artillery,* 59–60, 83–84; Hughes, *British Smoothbore Artillery,* 68–70; Hicks, *French Military Weapons,* 144–45; Dastrup, *King of Battle,* 66.

11. Peterson, *Round Shot and Rammers,* 66; Thaddeus Kosciusko, *Maneuver of Horse Artillery,* trans. by Jonathan Williams (New York: Campbell and Mitchell, 1808), 5; Hughes, *British Smoothbore Artillery,* 48.

12. Hughes, *British Smoothbore Artillery,* 103; Christopher Bellamy, *Red God of War: Soviet Artillery and Rocket Forces* (London: Brassey's Defence Publishers, 1986), 18–20; Rothenberg, *The Art of Warfare in the Age of Napoleon,* 74–80.

13. Rothenberg, *The Art of Warfare in the Age of Napoleon,* 74, 75, 107; Lynn, *The Bayonets of the Republic,* 209–11; Strachan, *European Armies and the Conduct of War,* 42–43; Quimby, *The Background of Napoleonic Warfare,* 296–97.

14. Rothenberg, *The Art of Warfare in the Age of Napoleon,* 77; Philip J. Haythornthwaite, *Weapons and Equipment of the Napoleonic Wars* (Poole, England: Blandford Press, 1979), 75–81; Kenneth Brookes, *Battle Thunder: The Story of Britain's Artillery*

(Reading, England: Osprey Publishing Ltd., 1973), 52–53; Hughes, *British Smoothbore Artillery,* 71.

15. Rothenberg, *The Art of Warfare in the Age of Napoleon,* 144, 181, 193, 201–2.

16. Dastrup, *King of Battle,* 47–49.

17. Rothenberg, *The Art of Warfare in the Age of Napoleon,* 77; Manucy, *Artillery through the Ages,* 69; Peterson, *Round Shot and Rammers,* 80; O.F.G. Hogg, *Artillery: Its Origin, Heyday, and Decline* (Hamden, CT: Archon Books, 1970), 179–83; Haythornthwaite, *Weapons and Equipment of the Napoleonic Wars,* 60.

18. Hughes, *British Smoothbore Artillery,* 56.

19. McNeill, *The Pursuit of Power,* 221; Hogg, *Artillery,* 247–50; Rogers, *A History of Artillery,* 194–95; Bernard and Fawn Brodie, *From Crossbow to H-bomb* (Bloomington, IN: Indiana University Press, 1982), 127; Frank H. Winter, *The First Golden Age of Rocketry,* (Washington, DC: Smithsonian Institution Press, 1990), xiv, 6, 81–141.

20. Brodie, *From Crossbow to H-Bomb,* 127; Winter, *The First Golden Age of Rocketry,* xiv, 6, 81–144.

21. Brodie, *From Crossbow to H-Bomb,* 131–32; Frank E. Comparato, *Age of Great Guns: Cannon Kings and Cannoneers Who Forged the Firepower of Artillery* (Harrisburg, PA: The Stackpole Company, 1965), p. 62.

22. Peterson, *Round Shot and Rammers,* 107; Comparato, *Age of Great Guns,* 61–62; Warren Ripley, *Artillery and Ammunition of the Civil War* (New York: Van Nostrand Reinhold Company, 1970), 233; Brookes, *Battle Thunder,* 90–91.

23. Peterson, *Round Shot and Rammers,* 107; Comparato, *Age of Great Guns,* 325.

24. Comparato, *Age of Great Guns,* 16–18.

25. Comparato, *Age of Great Guns,* 15–17; McNeill, *The Pursuit of Power,* 231–36; Strachan, *European Armies and the Conduct of War,* 117; Brodie, *From Crossbow to H-Bomb,* 132.

26. Comparato, *Age of Great Guns,* 18; James E. Hicks, *Notes on U.S. Ordnance: 1776–1941* (Green Farm, CT: Modern Books and Crafts, Inc., 1971), 139–40.

27. Stevens, *Artillery through the Ages,* 60; Rogers, *A History of Artillery,* 93–95; Ripley, *Artillery and Ammunition of the Civil War,* 366, 372; Curt Johnson, *Artillery: The Big Guns Go to War* (London: Octopus Books, 1975), 9.

28. William McElwee, *The Art of War: Waterloo to Mons* (Bloomington, IN: Indiana University Press, 1974), 124.

29. Trevor N. Dupuy, *The Evolution of Weapons and Warfare* (New York: Bobbs-Merrill, 1980), 174, 192.

30. McNeill, *The Pursuit of Power,* 237.

31. Ripley, *Artillery and Ammunition of the Civil War,* 78–79.

32. McNeill, *The Pursuit of Power,* 238; Brodie, *From Crossbow to H-Bomb,* 139–43; Jobe, *Guns,* 141–42.

33. Alexander L. Holley, *A Treatise of Ordnance and Armor* (New York: D. Van Nostrand, 1865), 50–53.

34. Peterson, *Round Shot and Rammers,* 110–12.

35. Dastrup, *King of Battle,* 82; Comparato, *Age of Great Guns,* 115; Johnson, *Artillery,* 10.

36. Johnson, *Artillery,* 18–20.

37. Johnson, *Artillery,* 18; Bellamy, *Red God of War,* 16, 28.

3

Introduction of Modern Field Artillery, 1866–1918

During the last four decades of the nineteenth century and the first two decades of the twentieth century, a technological revolution ended the age of smoothbore field artillery and direct fire. Armies adopted steel breechloaders with ranges that were dramatically greater than those of smoothbore muzzleloaders, recoil systems, high-explosive propellants, motor vehicles as prime movers, and indirect fire.

LAYING A FOUNDATION

Searching for a strong metal that would hold rifling better than wrought iron or bronze, the Prussians turned to steel during the 1860s. In 1866 the Prussians entered the Austro-Prussian War with a mixture of old and new field artillery. Each Prussian corps had sixteen batteries, six of them equipped with smoothbore muzzleloaders and the rest armed with Alfred Krupp's (q.v.) rifled steel breechloaders.In contrast, the Austrians supplied their army with rifled muzzleloaders. Although Prussian field artillery was superior, incompetent leadership, obsolete tactics, and inappropriate organization caused it to perform less effectively than desired. Equally important, Krupp's steel breechloaders leaked gases at the breech with some of them even exploding on the battlefield, as many wrought-iron breechloaders of the time were also doing. In light of the Prussians' technical troubles with their steel breechloaders, most European and American armies discounted them and persisted in furnishing their batteries with bronze or wrought-iron smoothbore and rifled muzzleloaders.[1]

Strongly disappointed with their field artillery's performance during the war with the Austrians, the Prussians instituted far-reaching reforms during the remaining years of the 1860s to eliminate tactical, organizational, and technical deficiencies. Tactically, the Prussians decided to push their guns to the front as rapidly as possible to engage the enemy's field artillery without coming within range of small arms fire, thereby obtaining fire superiority quickly. To implement this aggressive tactic the Prussians abolished the artillery reserve to deter commanders from holding back a portion of their guns and positioned their field guns in the center of the corps front to mass fire. Because rifled breechloaders had ranges of around 4,000 yards, field artillerymen did not have to assemble their guns in huge batteries or maintain an artillery reserve to mass fire as their predecessors had to do with their smoothbores, which had effective ranges of approximately 1,000 yards. To mass fire on a target, guns crews merely had to shift the direction of fire. In the meantime, Krupp perfected a method for producing flawless steel and developed a gas-tight steel breechloader with a higher rate of fire (five to six rounds a minute) than any field gun in use.[2]

In the Franco-Prussian War of 1870–1871, the Prussians employed their new field artillery, tactics, and organization for the first time. Although the French army under Napoleon III (q.v.) was the first to adopt rifled field artillery, and although its gun crews were well trained, an infatuation with the mitrailleuse, an early machine gun, and self-satisfaction with French accomplishments in the Crimean and Italian wars of the 1850s discouraged the French from rearming their field artillery with the new technology, even though it was readily available. The army remained equipped with bronze muzzleloaders and wrought-iron breechloaders, and consequently reliant upon Napoleonic artillery tactics and the artillery reserve. This situation left the French vulnerable to superior Prussian technology, organization, and tactics. During the war, Prussian gun crews frequently attacked French guns at ranges of 4,000 yards, effectively kept French batteries out of range with massed fire, aggressively pushed their pieces forward with the infantry advance as required, repeatedly moved within range of rifled muskets, and played a major role in the decisive Prussian victory, complementing the efficient work of the general staff. Simply stated, Prussian field artillery outclassed French artillery.[3]

Although some Krupp steel breechloaders failed at one time or another during combat, their general supremacy over French field artillery finally convinced Europeans to adopt steel breechloaders and led to an arms race in which European armies rushed to obtain the latest field artillery technology, which was constantly changing during the last decades of the nineteenth century. Even the most efficient state arsenals after the Franco-Prussian War of 1870–1871 had to purchase steel field artillery from private manufacturers such as the Krupps in Germany and Sir William Armstrong (q.v.) in Great Britain to keep up with foreign competitors; state arsenals were not equipped to produce steel artillery because the costs were politically prohibitive.[4] As a result of the combination of private manufacturers and the state arsenals, steel breechloaders dominated inventories by the 1890s.[5]

After finding a satisfactory breech-loading mechanism during the 1860s that was repeatedly improved upon during the remaining years of the century, Europeans worked to develop suitable recoil systems that made their steel breech-loaders even more effective and lethal. For years Europeans had searched for a practical way to return the cannon back into battery after the recoil had pushed it out of its firing position, a laborious task that was necessary because the cannon tube was mounted rigidly to the carriage. For example, British gun crews at the Battle of Waterloo became so exhausted from manhandling their guns back into battery that they started firing them from wherever the recoil had stopped. By the time the battle was over, some of the crews were firing from the wagon lines because the pieces had recoiled that far back.[6]

To eliminate the recoil that not only exhausted gun crews but also restricted the rate of fire to about two to three rounds a minute, inventors turned to the science of hydraulics, which was just emerging during the middle years of the nineteenth century. After successfully attaching hydraulic cylinders to coast artillery guns in the 1860s, inventors began mounting them on field carriages after developing the technology to produce small cylinders. One method involved employing a hydraulic buffer. The gun tube was mounted on a slide or cradle that was attached to the gun carriage and permitted the tube to move freely back and forth. When the cannon was fired, the gun tube recoiled down the cradle until it struck the buffer piston rod. The impact drove the rod into the buffer cylinder, where the piston head was forced against a body of water to absorb the recoil. Unfortunately, this only arrested the movement of the gun with less violence than before and did nothing to return the tube back into battery.[7]

About the same time, Woolwich designers in Great Britain introduced a more sophisticated recoil system. The system consisted of a cylinder of oil with a piston having a rod attached that was connected to the gun tube. When the gun recoiled, the tube slid down the cradle and forced oil through the hole in the piston head into a reservoir to slow down the recoil. Because the system lacked a means of automatically returning the tube back into battery, the gun crew had to manhandle the tube with handspikes down the cradle back into battery. Notwithstanding this undesirable characteristic, the British system laid the foundation for other systems to build upon.[8]

Early in the 1870s, Alfred Krupp took the next step. He mounted a cradle on a carriage to permit the artillery tube to slide back and forth with ease and connected the tube to a piston that was inside of a hydraulic cylinder. When the cannon recoiled upon firing, the piston was pulled through the hydraulic cylinder to absorb the recoil. This action compressed the spring inside the cylinder. When the recoil had stopped, the string pushed the tube back into battery. Although Krupp's system, which was introduced in 1873, was successful, it still allowed the carriage to shift on the ground and forced the crew to manhandle the piece back into battery before relaying the firing. On the basis of Krupp's accomplishment, most European field artillery manufacturers introduced hydraulic cylinders with huge, coarse springs inside, which had the tendency to choke off the recoil sharply and slam the gun back into battery, and had to resort to trail spades and wheel brakes to help reduce the recoil. Field guns with such recoil systems, often called short recoil systems because they recoiled about four to six inches on their carriages, were quickly adopted throughout Europe. Although the short recoil system did not prevent the gun and carriage from rolling backward on the ground, it was a vast improvement over guns that lacked a recoil system.[9]

Although the Germans dismissed Konrad Haussner's patented invention of a long recoil system that recoiled about one yard on the carriage, was smoother and faster than the existing short systems, and promised to make short recoil systems obsolete, the French did not and worked to develop one of their own. After years of effort in the 1890s, they took the lead by developing a long recoil system in their M1897 75mm field gun, based upon the work of Captains Émile Rimaihlo and Sainte-Claire Deville. This recoil system was composed of two parallel cylinders connected at the breech by a series of valves and a diaphragm, with the upper cylinder containing hydraulic fluid and a piston attached to a rod. In the lower cylinder a floating piston separated the hydraulic fluid from pressurized air. When the tube recoiled upon firing, it pulled the piston through the upper cylinder and pushed the fluid into the lower cylinder to slow down the tube's rearward movement. The movement of the fluid into the lower cylinder compressed the pressurized air. At the end of the recoil, the compressed air expanded to force the fluid back into the upper cylinder to propel the tube back into battery. The system was so sound that the cannon recoiled and returned to its firing position, while the carriage remained stationary to permit the gun tube to stay aligned on the target. Using the M1897's recoil system, which established the standard for other field guns to imitate, a well-trained French gun crew could, in a emergency, fire up to thirty rounds of aimed fire a minute with fixed ammunition up to 8,000 yards. Within years the Germans, Russians, British, and Americans were adopting field guns with long recoil systems that were comparable to the French M1897 75mm gun.[10]

Effective breech-loading mechanisms and recoil systems made it practical to add shields to protect the crew from small arms fire. Because the crew did not have to jump clear of the recoiling carriage, they could remain behind the gun. The combination of the recoil system, breech-loading mechanism, and shield meant that fieldpieces could be moved close to the target if such action was warranted and pour murderous rates of fire on it with the crew being protected from hostile small arms fire.[11]

In the meantime, the introduction of smokeless powder helped complete the revolution in field artillery materiel. Prodded to find a better propellant than black powder, which upon firing emitted a cloud of white smoke that obscured the battlefield, Henri Braconnot of France invented a highly inflammable substance in 1832 by treating sawdust or cotton with nitric acid. Even though Braconnot's invention was highly volatile and unmanageable, it proved to be not only comparatively smokeless but also more powerful than black powder and encouraged Europeans to continue experimenting with nitrogen-based compounds over the next several decades. Although a number of inventors improved upon Braconnot's work, a French chemist, Paul Vieille, developed the first dependable smokeless powder, called nitrocellulose (a combination of nitric acid and wood pulp or cotton), for military use in 1884. Building upon Vieille's invention, Alfred Nobel of Sweden combined nitrocellulose and nitroglycerine and patented his propellant as ballistite in 1888. Shortly afterward, the British introduced cordite, a mixture of nitrocellulose and nitroglycerine. These high explosives neither gave off smoke upon being fired nor obscured the battlefield.[18]

Two additional critical properties of smokeless powder magnified the performance of field artillery. While black powder produced high breech pressures, burned rapidly, and drove the projectile only a short distance down the cannon tube, smokeless powder burned more slowly and propelled the round longer in the tube to generate greater muzzle velocity, range, and power.[13] These characteristics permitted gun manufacturers to introduce longer tubes to take advantage of the longer burning time of smokeless powder and simultaneously reduce the size and weight of the cannon because the metal did not have to be as thick as it had been with black powder to contain the explosion. As a result of these developments, fieldpieces that appeared in the 1890s were lighter and more mobile than their predecessors. Equally important, the absence of smoke improved the rate of fire because gun crews did not have to wait for the smoke to clear off the battlefield before firing again. Aware of smokeless powder's desirable attributes, European armies enthusiastically adopted it in the 1890s.[14]

Concurrently, Europeans adopted high explosives as bursting charges for their ammunition. Although the Germans experimented with dynamite in the 1870s, the explosive was too stable for detonation by the simple fuzes of the day. Through the work of Vieille and his associate, Émile Sarrau, the French learned in 1886 to control picric acid, a highly explosive substance. The combination of picric acid and nitrocellulose was patented as melinite and was later accepted for French service as a shell filler. Soon, other European countries adopted similar explosives to detonate artillery rounds. Unlike black powder shells, which fragmented into five or six pieces upon detonation, high-explosive shells shattered into approximately one thousand splinters and were deadly against troops in the open.[15]

In the 1890s the introduction of metallic cartridge cases further improved the effectiveness of field artillery. During the decade, metallic cases to hold the charge and primer replaced cloth cartridge cases and united the projectile and charge into one unit. This arrangement simplified loading and firing because gunners no longer had to load the powder and projectile separately, while the metal case helped seal the breech by serving as an obturator, contained the primer to detonate the charge, and protected the charge against weather.[16]

High explosives, metallic cartridges, and long recoil systems led to the development of quick-fire (rapid-fire) field artillery in the 1890s. Together, these inventions increased the rates of fire and caused ranges to soar simultaneously. By the mid–1890s the average European fieldpiece could fire up to eight rounds of aimed fire a minute at ranges of 7,000 yards or more, with the French M1897 75mm gun firing a 12-pound projectile over 8,000 yards. This meant that one rapid-fire gun had the firepower of a battery of smoothbores without impairing accuracy or imposing extra exertion on the gun crew.[17]

Prompted by the greater ranges, European armies abolished the artillery reserve during the latter years of the nineteenth century as the Prussians had done early in the 1870s. Because field artillerymen could shift fire around the battlefield to mass on a target with relative ease, they no longer required the reserve. To facilitate massing fire and maintaining command and control, Major (later Sir) Edward S. May, a leading British field artillery intellectual of the time, wrote that all of the batteries had to be welded together into one tactical unit and directed by one person to achieve the desired results. In view of this thinking and the necessity of massing fire, most European armies attached their field artillery to the division and often the corps, formed their batteries into a battalion or a similar organization, and placed their field artillery under one commander. Influenced by the capabilities of the new technology as well as by the Prussians' aggressive field artillery tactics and effective general staff work during the Franco-Prussian War of 1870–1871, Europeans concurrently taught that all available guns should be firing at the beginning of the battle and not held back as a tactical reserve to be thrown into the battle later.[18]

Technology introduced during the last decades of the nineteenth century profoundly improved the performance, capabilities, and lethality of field artillery. As the century came to a close, field artillery was in a position to recapture the dominance that it once had on the battlefield. Even so, one major problem had to be solved before armies could harness the long ranges of their breech-loading field artillery.

FROM DIRECT TO INDIRECT FIRE

As profound as these organizational and tactical reforms were, they still fell short of exploiting the new field guns that were being introduced during the last decades of the nineteenth century.[19] As early as the 1870s, Europeans had recognized the dilemma posed by the long ranges (4,000 yards) of the new field artillery and small arms technology, and had initiated searches for a practical method of hitting a target beyond human eyesight, which was about 1,700 yards, and protecting the batteries from small arms fire. Because of their experience during the Franco-Prussian War, in which their batteries had suffered tremendously when they were advanced to within small arms range, the Prussians were the first to advocate concealing the batteries out of range of small arms to take advantage of their fieldpieces' superior ranges, even if this meant blocking the gun crew's view of the target. Although concealing the battery would protect the guns from rifled small arms fire, machine guns, and counterbattery work, it would simultaneously prevent the crew from seeing the target and hitting it with direct fire techniques. Consequently, the desire to hide batteries and exploit their guns' ranges compelled Prussian field artillerymen to espouse indirect fire, which had been practiced for some time by siege artillery. High-ranking German officers, however, pointed out that the gun crew had to see every shot and discounted indirect fire because it was too slow and imprecise for a mobile battlefield.[20]

As the Germans were hard at work to find a suitable method of indirect fire, a Russian army officer, Lieutenant Colonel Karl G. Guk (q.v.), published a book in 1882 entitled *The Covered Fire of Field Artillery,* in which he described the use of an aiming point, observer, and compass to hit an unseen target through indirect fire. The sight of one gun of a battery was laid on the aiming point, such as a stake or steeple, that could be easily seen by the forward observer and the gun crew. The forward observer then measured the angle formed by the aiming point, the gun, and the target and then relayed the information to the gun crew. The crew subsequently traversed the base piece and other guns, allowing for intervals, the requisite number of degrees to bring the guns onto the target. Range was estimated and adjusted by the forward observer. If shells landed to the right or left of the target, the forward observer corrected the deflection. Within years other European armies adopted this method of indirect fire because it was simple but effective for hitting targets beyond human eyesight and for protecting the guns from hostile artillery and small arms fire.[21]

The appearance of the French 75mm field gun reaffirmed the necessity of hiding batteries to protect them and employing indirect fire. In 1897, progressive field artillerymen in the French army pointed out that a plainly visible battery could easily be destroyed by counterbattery fire and that high terrain, the traditional location for field batteries during the age of smoothbores, was no longer an ideal spot. Agreeing, dynamic German field artillerymen, such as Prince Kraft zu Hohenlohe-Ingelfingen (q.v.), enthusiastically urged the adoption of indirect fire and avoided direct fire unless they were forced to do so. Later in 1897, the Germans, who advocated the new technique of fire direction, noted that the increased ranges, the precision of modern fire, and the introduction of smokeless powder made utilizing natural and artificial cover imperative. The field artillery had to keep out of sight as long as possible to be useful and had to employ indirect fire in order to survive on the battlefield.[22]

Interestingly, indirect fire, though one of the most pivotal developments in field artillery, still had limitations. With indirect fire field artillery had to rely upon the forward observer, who was tenuously linked to the guns with a field telephone or other means of communication, to locate targets and adjust fire. As a result, indirect fire depended upon good communications between the forward observer and the battery. Without good communications the batteries would be ineffectual because they would not know when and where to fire in support of the other combat arms.[23]

The adoption of indirect fire simultaneously compelled changes to be made in battery operations. During the era of direct fire, each gun crew fired its piece independently for the most part, even though the battery commander could have them fire in unison if necessary. With the advent of indirect fire, the observer aimed the battery's guns to make the battery a firing unit. Except when direct fire was employed, the independent aiming of each piece by its gun crew disappeared.[24]

Even though technology and techniques existed to employ indirect fire at the close of the nineteenth century, many field artillerymen did not readily embrace it. Direct fire was faster and easier than indirect fire and boosted the morale of the cavalry and infantry because the guns were on line with them during the battle. Moreover, conservative field artillerymen of the time felt that it was not gentlemanly to fire from protected positions and that the most attractive feature of field artillery was its ability to rush into battle, move within range of infantry weapons, fire a few rounds, and then rapidly move to another position as Napoleon's artillery had done earlier in the nineteenth century. Consequently, many army officers preferred direct fire even though it was obsolete and dangerous for the battery, and discounted indirect fire.[25]

The Russo-Japanese War of 1904–1905 quickly demonstrated the futility of direct fire with the new field artillery technology. Early in the war, the Japanese employed indirect fire techniques that were patterned after German methods to silence Russian batteries placed in the open for direct fire, while Russian gun crews, although they had modern rapid-fire pieces that were superior to their opponent's field artillery and were trained in indirect fire techniques, had problems engaging concealed Japanese batteries. In light of this devastating experience and the Japanese successes, the Russians converted to indirect fire during the middle of the war to protect their field guns from intense counterbattery fire.[26]

After observing the Russo-Japanese War, European and American armies quickly altered their attitudes about indirect fire. Between 1905 and 1914, armies introduced quick-fire fieldpieces similar to the French 75mm gun. The German 77mm, the British 13-pounder (3-inch bore) and 18-pounder (3.3-inch bore), the American 3-inch, and the Russian 76.2mm field guns, with their flat trajectories and high rates of fire (twenty to thirty rounds a minute in an emergency), were suitable for indirect fire, and field artillerymen were learning indirect fire techniques.[27]

At the same time, the Russo-Japanese War reinforced lessons of the Russo-Turkish War of 1877–1878 indicating that howitzers were necessary to attack enemy forces entrenched to protect themselves from hostile artillery and small arms fire, and to hit targets concealed on reverse slopes. In light of this, European and American armies integrated howitzers into their divisional artillery after they had minimized the weapon's value for so many years. The British introduced eighteen howitzers in their division. The Germans placed twelve howitzers in their division, while the Russians equipped their division with three or four. Caught in the middle of the debate over the appropriate size of howitzers, the French army did not have any in their division when World War I broke out in 1914, even though maneuvers in 1913 and trials with an experimental 105mm howitzer indicated the effectiveness of howitzers as a part of the division's artillery.[28]

Although armies adopted indirect fire and new field artillery weapons, they still differed over suitable tactics. Influenced by their formidable 75mm field gun, the French stressed mobility at the cost of firepower and preferred short bursts of fire of five or six rounds to force the enemy to seek cover, allowing the infantry to advance. Equally important, the French still maintained the artillery reserve even though it was antiquated with the advent of long-range rifled breechloaders. Because of their experience in the Boer War and observer reports of the Russo-Japanese War, the British also emphasized short but violent bursts of fire from concealed batteries to neutralize the enemy. Yet, they also taught that the guns should be moved forward to positions within small arms range to furnish direct fire if necessary, even though recent wars indicated that such a practice was suicidal. Although the Germans taught that artillery fire should neutralize and destroy through indirect fire and had the most balanced tactics of the European powers, they also warned field artillerymen that they would have to come out in the open for direct fire to press home the attack during the decisive stage of the battle if required. American tactics, meanwhile, closely resembled their French and British counterparts. Indirect fire was accepted, but European and American armies were not willing to abandon direct fire totally, even though the Russo-Japanese War of 1904–1905 had revealed its futility. The old ways of smoothbore artillery still persisted.[29]

Hence, on the eve of World War I, European and American armies were geared to fight a war of movement, for the most part, in which mobility was the most desirable trait for the field artillery, although the Russo-Turkish War had bolstered the lessons of the American Civil War of 1861–1865. Faced with intense small arms and field artillery fire, the infantry would dig in for protection. Entrenching had the potential of restricting and even eliminating mobility; it stressed the desirability of field howitzers, because their curved trajectories could hit inside trenches, and it underscored the requirement for even more firepower to destroy fortifications.[30]

The revolution in field artillery technology and fire direction techniques after 1866 indicated that a new era had opened. As doctrine, tactics, and organization revealed, however, many field artillerymen encountered the new age with some ambivalence and reluctance by preserving tactics and techniques that were ill-suited for the new technology. The indirect fire was intimidating because it was more complicated than direct fire and because it represented a total break with the days of direct fire.

THE NEW TECHNOLOGY AND WORLD WAR I

World War I offered the first large-scale opportunity to test the new technology and indirect fire under intense combat conditions. Following the August 1914 offensive, combat on the western front settled down into trench warfare that caused the field artillery to adjust to new, unforeseen circumstances. To overcome the trenches that had resulted from the superiority of firepower over mobility, Europeans increased the number of heavy artillery pieces (often siege and railway artillery) in relation to the number of light field guns. For example, the French had almost one thousand 75mm gun batteries and fifty heavy batteries (batteries of artillery pieces that were larger than the 75mm gun) in August 1914. Four years later in 1918, the French had 1,014 heavy batteries, while their number of light batteries had remained constant. The use of heavy siege artillery as field artillery blurred the distinction between the two and prompted many field artillerymen to question the rationale of distinguishing between siege and field artillery. Also, by 1915 long, exhausting bombardments to destroy elaborate trench defenses often lasted for days and replaced rapid bursts of fire from light field guns.[31]

Besides bringing prewar field artillery tactics to an abrupt halt and highlighting the necessity of heavier field guns, trench warfare simultaneously accented the difficulties of providing effective counterbattery work with observed indirect fire, which was in its infancy. Since guns on the offensive had to silence enemy artillery for the infantry to advance, Europeans devised various methods of detecting concealed batteries, besides employing aircraft to find them. Early in the war, they introduced sound ranging. It was composed of an arc of microphone pickup stations to record the sounds of the guns and calculate their direction and distance. Instruments of the time could locate a battery to within 50 to 60 yards. By the end of the war, sound ranging had found more guns than any other method of target acquisition. In the meantime, Europeans developed flash ranging to detect gun positions by picking up their flashes. Before long, however, field artillerymen were devising ways to disguise the sound and flashes of their guns to make counterbattery fire even more challenging and demanding because of the difficulty of locating hostile batteries.[32]

Whether for counterbattery work or close support, the artillery barrages produced smoke and dust, hindering visual communications, and cut telephone lines, further disrupting communications between the observer and the battery. These problems, along with undependable field telephones and the unreliability of other communications technology of the time, led armies to find other means than observed indirect fire to engage the enemy with artillery fire. By 1916 the French had resorted to unobserved or map firing. Rather than relying upon a forward observer to identify targets for engaging, unobserved indirect fire employed maps of enemy positions produced from aerial photographs or information from sound-and-flash ranging. Once the targets had been pinpointed on the map, gun crews then calculated the range and direction of the target from their batteries, incorporated weather data into their calculations because wind speed and direction, humidity, temperature, and barometric pressure influenced the projectile's accuracy, and opened fire. Although observed indirect fire could be effective on occasion, unobserved indirect fire, even though it had been known prior to the war but not practiced extensively, became the favorite technique of most combatants by 1917 for attacking targets. It was easier to conduct than observed indirect fire and was not dependent upon vulnerable communications networks.[33]

The same factors that led to the general acceptance of unobserved indirect fire influenced the British to develop the rolling barrage. Soon adopted by the other combatants and favored over observed indirect fire, the rolling barrage laid down a curtain of fire about 300 to 400 yards in front of the infantry advance to compel the enemy to seek cover and moved forward at a planned rate of speed. Frequently, the barrage left the infantry behind because craters caused by the shelling, barbed wire entanglements, and other manmade and natural obstacles slowed down the infantry's forward movement. If the rolling barrage moved too far ahead of the infantry, the defense had time to recover after the shelling had passed overhead and could open with machine gun and rifle fire on the attacking infantry. Despite this problem, armies preferred the rolling barrage over observed indirect fire because it did not depend upon unreliable communications systems.[34]

As effective as rolling barrages and map fire were, the introduction of predicted fire in 1917 added another critical dimension to fire support that further discouraged the use of observed indirect fire. Although the British initially employed predicted fire (hitting a target without any ranging shots to achieve surprise) without much success during the Boer War in the 1890s, technical improvements in maps, survey, aerial photography, and meteorological data permitted accurate projection of where the projectile would land, reducing the requirement for registration, and made this type of fire a reality by the time that World War I broke out. As with the rolling barrage and unobserved fire, predicted fire did not hinge upon good communications between the forward observer and guns and between the field artillery and other combat arms to be effective, as observed indirect fire did.[35]

At Cambrai in November 1917, for example, the British used predicted fire effectively for the first time in combat. Because surprise was essential, British guns remained silent on their arrival in their battle positions, engaged in no artillery duel to neutralize enemy artillery, and did not bombard the hostile trenches prior to the battle. The operation began with intense predicted fire in front of the advancing tanks, with the infantry following to mop up the surprised defenders. The combination of tanks, infantry, and new artillery techniques paved the way for a brilliant start. However, the German army recovered, and the British offensive soon stalled. Even so, the British ability to hit targets efficiently without registration was a significant breakthrough in artillery tactics and was soon adopted by other armies.[36]

Thus, early in the war, combatants abandoned observed indirect fire, with its emphasis upon shifting fire rapidly around the battlefield to engage targets of opportunity as they appeared to allow the infantry to advance. The state of communications technology and battlefield conditions prohibited effective command and control, upon which observed indirect fire was dependent, and close support with this form of fire rarely performed well. With unobserved indirect fire, rolling barrages, and predicted fire, command and control was less critical, because fire plans were rigid and because the other combat arms were expected to adhere to them. As long as action went as planned, the field artillery could furnish fire support to the advancing infantry with rolling barrages, map firing, and predicted fire. When impromptu fighting broke out that deviated from the plans, the inability to communicate efficiently forced the field artillery to resort to blind shelling of points on maps or to stop firing because of the fear of hitting its own infantry. The prewar vision of boldly shifting observed indirect fire around the battlefield to supply close support simply was not a reality because of the state of communications technology and battlefield conditions.[37]

The inability to furnish effective close support with observed indirect fire, as well as the difficulties of moving the guns forward over chewed-up ground to keep up with the infantry advance, also led to the adoption of the accompanying gun. First employed by the French and soon adopted by other combatants on the western front, these guns were light artillery pieces that could be drawn by infantrymen, who could devote all of their attention to the needs of the infantry and could shell targets quickly and effectively with direct fire. Although field artillerymen opposed the accompanying gun concept because it threatened to minimize their contribution to the combined arms team and reinforced the ineffectiveness of close support from the field artillery, they had to admit that the accompanying gun concept was unfortunately necessary until close support from field guns could be improved. Yet, the accompanying gun never supplanted field artillery as many field artillerymen feared, because most of the guns were 37mm pieces and lacked sufficient power to do serious damage to sophisticated defenses. Even so, the accompanying gun concept illustrated the extent to which the infantry would go to have effective close support when it encountered stiff resistance because of the breakdown in infantry-artillery cooperation brought on by the difficulties of employing observed and unobserved indirect fire with existing technology and techniques.[38]

Just as the war forced changes in tactics and highlighted the weakness of indirect fire, it simultaneously compelled the combatants to reexamine field artillery ammunition. Long considered the standard projectile, shrapnel performed effectively against troops in the open. Shrapnel was a projectile filled with iron balls that were shot forward in a cone-shaped pattern just as the shell arrived over the target by a preset time fuze, whereas high-explosive steel shell burst and scattered tiny steel fragments. During the war, armies found shrapnel wanting when it was used against entrenched troops because it lacked sufficient power to damage sophisticated fortifications. This encouraged the Europeans to turn to high-explosive steel shell, which had been around for years, as their preferred munition because it was more powerful and could tear apart entrenchments.[39]

In the meantime, the Europeans introduced gas-filled artillery shells. Unwilling to accept the indecisiveness of the war, the Germans carried out the first major gas attack, although the French had experimented with various noxious gases on a small scale at the end of 1914. In April 1915 in the Ypres sector, the Germans released lethal chlorine gas from cylinders. Although the escaping gas caused the enemy to panic, the Germans could not take advantage of their gas offensive because they had not known that the gas would work so well and because they lacked the reserves to exploit the break in the enemy lines. Shortly afterward, the French and British started discharging gas from cylinders and allowing the prevailing winds to carry it over to the Germans. Unable to depend upon the prevailing winds to blow the gas over enemy positions, the Germans then turned to gas-filled artillery shells in 1916 to hit enemy positions. In response, the British and French obtained their own gas artillery shells. In 1917 combatants on the western front expanded their use of gas shells to counterbattery fire because they could neutralize batteries faster with these shells than with conventional artillery rounds. In light of this development, French, British, and German artillery commanders counted heavily upon gas rounds for counterbattery work by 1918, and the number of shells fired was restricted only by their availability.[40]

Chemical artillery rounds certainly did not overshadow the introduction of motor vehicles as prime movers for the field artillery. With the appearance of motor vehicles and armored tanks during World War I, armies pushed to endow their field guns with the same mobility. One option involved using motor traction to tow the guns, while the other consisted of developing self-propelled artillery. Either case meant replacing the horse as the prime mover and forsaking a long-standing tradition in the process. Although many field artillerymen had visualized motor traction for their guns prior to 1914, the Great War prodded them to motorize their field artillery sooner than anticipated because of the development of heavy artillery to batter down fortifications, the large number of horses being killed on the battlefield, and the huge amounts of fodder being consumed by the horses. One of the earliest attempts at developing self-propelled artillery occurred when the British modified a tank chassis to carry a 60-pounder gun. Even though the British attempt failed to work because of technological difficulties, the French placed light field guns on caterpillar tractors to get them as far to the front as possible for close support and to complement motor-drawn medium and heavy artillery. By 1918 the French army had eight types of self-propelled artillery, making them a leader in the field.[41]

Although motor-drawn and self-propelled artillery lacked sufficient mobility and did not contribute much in combat, they constituted a critical milestone. Through World War I field artillery had been horse-drawn. On the basis of the initial successes of motorized (towed and self-propelled) artillery, field artillerymen acknowledged the potential of tractors and trucks as prime movers for their guns and howitzers. Yet, many remained skeptical because of unreliable gasoline engines, the requirement for spare parts, and the lack of trained mechanics.[42]

Although the technological revolution of 1866 to 1918 had improved the field artillery's lethality and had opened a new age, it simultaneously hindered combined arms warfare. The new technology forced the field artillery to rely upon indirect fire to survive on the battlefield. As the war illustrated, observed indirect fire was ineffective for the most part and forced reliance upon rigid map fire, the rolling barrage, and predicted fire. Until communications systems could be improved, the field artillery would have difficulties providing effective and flexible close support on a mobile battlefield, and combined arms warfare would be severely frustrated. Ultimately, World War I revealed the limitations of indirect fire spawned by the new technology and the necessity of further improvements before the field artillery could supply the close support required by the other combat arms.

NOTES

1. "Austrian and Prussian Armaments," *Army and Navy Journal,* 16 June 1866, 683–84; William Manchester, *The Arms of Krupp, 1587–1968* (Boston: Little, Brown, 1968), 100–101; "Breech-loading Ordnance," *Army and Navy Journal,* 26 June 1873, 738; Prince Kraft zu Hohenlohe-Ingelfingen, *Letters on Artillery,* trans. by Colonel N. L. Walford (London: Edward Stanford, 1898), 59–74; Arthur T. Coumbe, "Combined Arms Operations in the Franco-Prussian War of 1870–1871," *Field Artillery,* August 1988, 11; Curt Johnson, *Artillery: The Big Guns Go to War* (London: Octopus Books, 1975), 28.

2. *Army and Navy Journal,* 25 November 1870, 165; Heinrich Rohne, *The Progress of Modern Field Artillery,* trans. by M. M. Macomb (Washington, DC: Journal of the United States Infantry Association, 1908), 18–19; William McElwee, *The Art of War: Waterloo to Mons* (Bloomington, IN: Indiana University Press, 1974), 141–42; Johnson, *Artillery,* 28; Christopher D. Bellamy, "The Russian Artillery and the Origins of Indirect Fire: Part I," *Army Quarterly and Defense Journal,* April 1982, 212–13; Frank E. Comparato, *Age of Great Guns: Cannon Kings and Cannoneers Who Forged the Firepower of Artillery* (Harrisburg, PA: The Stackpole Company, 1965), 28, 104; W. J. Savoy, "The Evolution of the American Modern Light Gun," master's thesis, U.S. Army Command and General Staff College, 1978, 9; O.F.G. Hogg, *Artillery: Its Origins, Heyday, and Decline* (Hamden, CT: Archon Books, 1970), 135–41; Ian Hogg, *The Guns: 1914–1918* (New York: Ballantine Books 1971), 16–18.

3. Rohne, *The Progress of Modern Field Artillery,* 11–14; Robert S. Ballagh, Jr., "The Development of French Field Artillery Doctrine, 1870–1914: Technological Change and Tactics," master's thesis, Duke University, 1971, 2–7; Joseph Jobe, ed., *Guns: An Illustrated History of Artillery* (Greenwich, CT: New York Graphic Society, 1971), 150–51; Jonathan B.A. Bailey, *Field Artillery and Firepower* (Oxford, England: The Military Press, 1989), 116–17.

4. William H. McNeill, *The Pursuit of Power: Technology, Armed Force, and Society since* A.D. *1000* (Chicago: University of Chicago Press, 1982), 242; Jobe, *Guns,* 154.

5. McNeill, *The Pursuit of Power,* 242, 255–56.

6. Hogg, *The Guns,* 11–12.

7. John Batchelor and Ian Hogg, *Artillery* (New York: Charles Scribner's Sons, 1972), 13.

8. Ibid., 9, 13, 14.

9. Comparato, *Age of Great Guns,* 32–38, 41, 321; Rohne, *The Progress of Modern Field Artillery,* 18–19; Batchelor and Hogg, *Artillery,* 14.

10. Hogg, *The Guns,* 12; Comparato, *Age of Great Guns,* 35–42; Shelford Bidwell and Dominick Graham, *Firepower: British Army Weapons and Theories of War, 1904–1945* (London: George Allen and Unwin, 1982), 8–9. Haussner, who worked for Krupp, first proposed his long recoil system in 1888 and even constructed a pilot model using his system. Tested at the Gruson proving ground, the gun performed poorly because the ground was so hard that the trail spade would not dig into it. In view of this performance, the German artillery board rejected Haussner's concept and continued supporting short recoil systems into the 1890s.

11. Comparato, *Age of Great Guns,* 32–38, 41, 321; Rohne, *The Progress of Modern Field Artillery,* 18–19; Batchelor and Hogg, *Artillery,* 14; Bidwell and Graham, *Firepower,* 9; Hogg, *The Guns,* 14.

12. Hogg, *The Guns,* 16–18.

13. Comparato, *Age of Great Guns,* 85.

14. A. B. Dyer, *Handbook for Light Artillery* (New York: John Wiley and Sons, 1898), 120–33; Comparato, *Age of Great Guns,* 39; Savoy, "The Evolution of the American Modern Light Field Gun," 10; Sisson C. Pratt, *Field Artillery: Its Equipment, Organization, and Tactics* (London: Kegan Paul, Trench, Truebner, and Company, 1902), 6–7; Hew Strachan, *European Armies and the Conduct of War* (London: George Allen and Unwin, 1983), 119; Rohne, *The Progress of Modern Field Artillery,* 14–15.

15. Comparato, *Age of Great Guns,* 82–95.

16. Rohne, *The Progress of Modern Field Artillery,* 16–17; Comparato, *Age of Great Guns,* 30.

17. F. C. Morgan, *Handbook of Artillery Materiel* (London: Clowes and Sons Ltd., 1899), 24; Ballagh, "The Development of French Field Artillery Doctrine, 1870–1914," 31–32; Rohné, *The Progress of Modern Field Artillery,* 20.

18. Sir Edward S. May, *Achievements of Field Artillery* (Woolwich: Royal Artillery Institution, 1893), 120, 161–62; Rohne, *The Progress of Modern Field Artillery,* 55–61, 71, 75, 76; Sir Edward S. May, *Field Artillery with the Other Arms* (London: Sampson, Low, Marston, and Company, 1898), 62–66, 92–98; Oliver S. Spaulding, *Notes on Field Artillery* (Leavenworth, KS: U.S. Cavalry Association, 1917), 38–44.

19. Bailey, *Field Artillery and Firepower,* 118–19.

20. Ibid., 119–20; Comparato, *Age of Great Guns,* 95–99, 41, 321; Christopher Bellamy, *Red God of War: Soviet Artillery and Rocket Forces* (London: Brassey's Defense Publishers, 1986), 28; Strachan, *European Armies and the Conduct of War,* 119; Bidwell and Graham, *Firepower,* 8, 11, 12; Robert H. Scales, Jr., *Firepower in Limited War* (Washington, DC: National Defense University Press, 1990), 6–7; Moriz von Reichold, "Indirect Fire," *Journal of the United States Artillery,* September-October 1897, 171–75; Ernest Hinds, trans., "Employment of Artillery Fire," *Journal of the United States Artillery,* September-October 1904, 156; Hinds translated a series of articles by Colonel Percin of the French field artillery written in 1899 and 1900 to explain the French method of indirect fire.

21. Comparato, *Age of Great Guns,* 95–99; Bellamy, *Red God of War,* 28; Strachan, *European Armies and the Conduct of War,* 119, von Reichold, "Indirect Fire," 171–75; Hinds, "Employment of Artillery Fire," 156; Bidwell and Graham, *Firepower,* 28–30; Bellamy, "The Russian Artillery and the Origins of Indirect Fire: Part I," 212–15.

22. Von Reichold, "Indirect Fire," 170; M. Wallut, trans., "Employment of Rapid Fire Artillery in the Field," *Journal of the United States Artillery,* 1905, 204; Comparato,

Age of Great Guns, 97; Bellamy, "The Russian Artillery and the Origins of Indirect Fire: Part I," 214.

23. Strachan, *European Armies and the Conduct of War,* 119.

24. Ibid.; Konrad F. Schreier, Jr., "The U.S. Army 3.2-inch Field Gun," *Military Collector and Historian,* Fall 1972, 77; Rogers Birnie, *Gun Making in the United States* (Washington, DC: Government Printing Office, 1905), 27, 33.

25. Boyd L. Dastrup, *King of Battle: A Branch History of the U.S. Army's Field Artillery* (Fort Monroe, VA: Office of the Command Historian, U.S. Army Training and Doctrine Command, 1992), 148–49.

26. Johnson, *Artillery,* 43–45; Charles Wheeler, *A Discussion of the Proposed Methods to Increase the Rapidity of Fire of Field Guns* (Washington, DC: Government Printing Office, 1901), 8; Christopher D. Bellamy, "The Russian Artillery and the Origins of Indirect Fire: Part II," *Army Quarterly and Defense Journal,* July 1982, 332–33.

27. Strachan, *European Armies and the Conduct of War,* 138–39; H. A. Bethel, *Modern Artillery in the Field* (London: Macmillan, 1911), 241; Batchelor and Hogg, *Artillery,* 16.

28. Bidwell and Graham, *Firepower,* 13; Strachan, *European Armies and the Conduct of War,* 139; Bethel, *Modern Artillery in the Field,* 2; Comparato, *Age of Great Guns,* 115–18.

29. Jonathan House, *Towards Combined Arms Warfare: A Survey of Tactics, Doctrine, and Organization in the Twentieth Century* (Fort Leavenworth, KS: Combat Studies Institute, U.S. Army Command and General Staff College, 1984), 16–17.

30. Ibid., 16–17; William McElwee, *The Art of War: Waterloo to Mons* (Bloomington, IN: Indiana University Press, 1974), 204–5; Bidwell and Graham, *Firepower,* 10–18; Bailey, *Field Artillery and Firepower,* 120–25; Johnson, *Artillery,* 35; Comparato, *Age of Great Guns,* 115–16; Kenneth Brookes, *Battle Thunder: The Story of Britain's Artillery* (Reading, England: Osprey Publishing Ltd., 1973), 138.

31. Bailey, *Field Artillery and Firepower,* 130–52; Shelford Bidwell, *Gunners at War: A Tactical Study of the Royal Artillery in the Twentieth Century* (London: Arms and Armour Press, 1970), 34.

32. Comparato, *Age of Great Guns,* 99; William R. Bursell, "American Sound Ranging in Four Wars," *Field Artillery Journal,* November-December 1981, 53.

33. Colonel Maitre, "Evolution of Ideas in the Employment of Artillery during the War," *Field Artillery Journal,* September-October 1922, 375, 378, 380; Alexander T. Jennette, "Mass Fire in World War I," *Field Artillery Journal,* May-June 1975, 40–43; Bidwell and Graham, *Firepower,* 18, 68; Bidwell, *Gunners at War,* 33–34; M. E. Locke, "Artillery in Europe," *Field Artillery Journal,* July-September 1917, 294–301.

34. Bidwell and Graham, *Firepower,* 83–85; Jennette, "Mass Fire in World War I," 45.

35. Bidwell and Graham, *Firepower,* 9, 10, 91; Batchelor and Hogg, *Artillery,* 19; Brookes, *Battle Thunder,* 142–43.

36. Bidwell and Graham, *Firepower,* 91–92, 134–35; Batchelor and Hogg, *Artillery,* 19; Hogg, *The Guns,* 133.

37. Bidwell and Graham, *Firepower,* 113–14, 196–97; Ian V. Hogg, *Artillery in Color: 1920–1963* (New York: Arco Publishing, 1980), 8; Walter W. Merrill, "The Employment of Artillery in the Present War," *Field Artillery Journal,* July-September 1916, 416–17.

38. Comparato, *Age of Great Guns,* 158–59.

39. Batchelor and Hogg, *Artillery,* 19; Bernard and Fawn Brodie, *From Crossbow to*

H-Bomb (Bloomington, IN: Indiana University Press, 1973), 191–92; Brookes, *Battle Thunder,* 142–43.

40. Charles E. Heller, *Chemical Warfare in World War I: The American Experience, 1917–1918* (Fort Leavenworth, KS: Combat Studies Institute, U.S. Army Command and General Staff College, 1984), 7–11, 18–19, 86; Comparato, *Age of Great Guns,* 45–46.

41. Bailey, *Field Artillery and Firepower,* 154; H.C.B. Rogers, *A History of Artillery* (Secaucus, NJ: The Citadel Press, 1975), 169–71; Bellamy, *Red God of War,* 156.

42. Hogg, *The Guns,* 149–52.

4

From Peace to Cold War, 1919–1991

During the seven decades following World War I technological advances revolutionized weapons, munitions, mobility, command, control, and communications, and target acquisition. These improvements gave the field artillery the ability to furnish effective close support with observed indirect fire, something that it had been unable to do during World War I.

BUILDING UPON THE PRECEDENTS OF WORLD WAR I

Even though they faced budgetary austerity caused by pacifism and the Great Depression, during the 1920s and 1930s European and American armies continued motorizing their field artillery, increasing firepower, and improving observed and unobserved indirect fire, albeit slowly. The argument for replacing horses with motor vehicles as prime movers grew even more compelling. Field artillerymen knew that employing horses restricted the weight and caliber of the fieldpieces because horses could pull only so much, that men riding into battle would be fresher than those who walked, that mechanical power required less manpower, and that fuel for motor vehicles was less bulky than fodder for horses. Although these reasons provided a convincing rationale for motorization, the declining horse population played an equally important role. With the escalating use of motor vehicles for civilian purposes, horse-breeding farms slowly disappeared in the 1920s and 1930s. As a result, mobilizing the field artillery along the lines of 1914–1918 would be impossible without the right kind of horse. Pressed with the need for a prime mover, field artillerymen searched for the proper motor vehicle. Those favoring tactical mobility championed track vehicles and rejected wheeled vehicles because the former offered greater cross-country mobility.[1]

Some field artillerymen even advocated portee artillery, which had been a French expedient during the war to overcome a draft animal shortage. With this type of field artillery, a conventional truck and a low-wheeled trailer transported the cannon, caisson, and prime mover (usually a caterpillar tractor) over the road at a high speed. In some instances, the gun was carried on the bed of the truck. At the gun's destination the gun crew dismounted the equipment from the trailer or truck by ramp for final cross-country positioning. For many field artillerymen, portee artillery held promise until about 1930, when opposition from conservatives in American and European armies discounted it because horse- and truck-drawn field guns were more maneuverable and less conspicuous.[2]

Interest in self-propelled artillery also soon waned. By 1922 postwar experiments in the United States had produced twelve different models. Of the twelve, those manufactured by the Holt Caterpillar Company and by Walter Christie attracted the most attention. Even though some self-propelled models were promising, the Americans abandoned self-propelled artillery because it was too difficult to maintain, too expensive, and relatively immobile. Equally important, self-propelled artillery's high silhouette presented an inviting target for counterbattery fire. In view of these unfavorable characteristics, the Americans channeled their energies toward developing tractors and trucks for towing field artillery after the middle of the 1920s. They were not alone. The Germans also suspended work on self-propelled artillery for the same reasons as the Americans.[3]

Other armies followed the American and German example. By the beginning of the 1930s, most countries were experimenting with towed artillery and had almost forgotten self-propelled artillery, because the latter was deemed less desirable. Equally important, if towed artillery's prime mover broke down, horses could pull the weapon if necessary to provide close support. With self-propelled artillery this was simply out of the question because of the weapon's weight.[4]

Before horses could be abandoned, however, manufacturers had to design suitable carriages for towing behind a motor vehicle and find satisfactory motor vehicles. Existing horse-drawn carriages had steel-rimmed, wooden wheels that could not be towed any faster than two to three miles per hour. In their efforts to convert to motor-drawn artillery, inventors modified horse-drawn carriages by equipping wheels with ball bearings, steel rims, and pneumatic tires for rapid movement as a temporary expedient until carriages specially designed for towing by a motor vehicle could be manufactured.[5]

As work progressed with carriages, Europeans and Americans experimented with different motor vehicles to determine the best. After over a decade of work, the Americans concluded early in the 1930s that the technology existed to make towed artillery a reality. This, coupled with the declining horse population in the United States, encouraged the American army to announce plans in 1933 to motorize 50 percent of its light artillery for the division to complement its medium and heavy artillery, which had already been motorized in the 1920s. Concurrently, the Europeans and Japanese converted a portion of their field guns to towed artillery by the end of the 1930s.[6]

With the improvements in motor vehicles, much of the skepticism about self-propelled artillery simultaneously disappeared in Europe and the United States during the 1930s. By the end of the 1930s, the Germans were manufacturing mobile, self-propelled 105mm howitzers. German accomplishments with their self-propelled artillery in Poland in 1939 and France in 1940 inspired less progressive nations to acquire their own. With the possibility of a German invasion after Dunkirk in 1940, the British pursued self-propelled artillery with greater vigor than ever before. Because of the requirement for mobile weapons that could be moved around rapidly to defend threatened areas along the coasts, the British mounted a 25-pounder (3.45-inch) field gun on a tank chassis to serve as a self-propelled piece. Shortly afterward, the Americans improvised their own self-propelled artillery when they installed a 105mm howitzer tube on a medium tank chassis in 1942 to stay abreast of German developments. Later, once their new 155mm towed gun had begun appearing, the Americans mounted serviceable World War I 155mm guns on converted tank chassis and fitted the combination with a bulldozer spade at the rear for dropping to the ground to give the weapon a solid base for firing.[7]

Like their American, German, and British counterparts, the Soviets also embraced self-propelled artillery. Although the Soviets employed self-propelled guns for direct fire to support infantry and armored forces and for antitank work, they also used them for indirect fire. By the end of World War II, self-propelled field artillery had become a common sight in combat and effectively supported armored forces because it was mobile, could be emplaced and displaced more rapidly than towed artillery, and offered protection from small arms fire for the gun crew.[8]

Even though self-propelled artillery vindicated its advocates during World War II, converting to towed and self-propelled artillery generated concern with some field artillerymen during the interwar years. Adopting motor vehicles as prime movers involved forsaking the tradition that was intimately associated with the horse and developing new organization, tactics, and equipment. For many field artillerymen, such as Major General Robert M. Danford, Chief of Field Artillery, U.S. Army, from 1938 to 1942, this created anxiety and even resistance because many were not totally convinced that motor vehicles were more dependable and better than horses off improved roads. In view of these fears and apprehensions, many field artillerymen did not want to give up the tried and proven for the unknown. Yet, the break with horse-drawn artillery accelerated during World War II, with the Americans grabbing the lead after the initial German successes of 1939–1941 had underscored motorized artillery's mobility and reliability and completed the revolution in field artillery transportation that had started during World War I.[9]

Notwithstanding the successes of towed and self-propelled artillery, many armies still retained horses and even mules as means of transportation during World War II. Pack artillery, which depended upon horses or mules as prime movers, was still being used, especially in mountainous terrain in Italy and in the China-Burma-India theater, where motor vehicles had difficulties traveling, while the Germans and Russians retained horse-drawn artillery as the bulk of their field artillery forces because of the cost of converting. For example, of the 153 German divisions that participated in Operation Barbarossa in 1941, 119 were horse-drawn.[10]

As armies were adopting motorized artillery, they simultaneously pushed to introduce more powerful fieldpieces. Interested in developing a modern field artillery system after depending upon the French and British for weapons during World War I, the American army convened the Westervelt Board of 1919 and other boards to determine future armament. In its report the Westervelt Board recommended furnishing a gun-and-howitzer combination for each echelon from division to army and producing new projectiles, fuzes, and motorized artillery, both towed and self-propelled. Like the Americans, the British and French examined their field artillery in light of the war experiences of 1914–1918 and saw the need for new artillery. However, a war surplus, the lack of funds, and public reaction in the United States and Europe against the carnage of the Great War slowed down developments in the 1920s and early 1930s, while Versailles Treaty provisions seriously limited German armament through 1933.[11]

In 1933 the Europeans and Americans crossed an important watershed with their field artillery. Because the Versailles Treaty had stripped the country of World War I field artillery, Germany standardized a new 88mm gun in 1934 that proved its versatility in the Spanish Civil War of 1936–1939, and replaced its 77mm field gun with a 105mm howitzer having a split-trail carriage for greater traverse and elevation and a range of 10,300 yards. Aroused by German rearmament after Hitler's accession to power in 1933, other European countries began seriously reequipping their armies with new field artillery. In 1933 the British initiated work on a 25-pounder (3.45-inch) fieldpiece with a range of 13,400 yards, which went into service in 1940–1941. Meanwhile, the French adopted a new 105mm howitzer during the latter years of the 1930s with a range comparable to that of the British 25-pounder. At the end of the 1930s, the Americans had an experimental towed 105mm howitzer with a range of 12,500 yards. As these countries worked to introduce new fieldpieces, the Soviets manufactured their own and had a higher proportion of artillery to their other combat arms than Western countries because of their propensity to rely upon firepower on the battlefield. All of these new fieldpieces that were adopted during the late 1930s or that were on the drawing boards on the eve of World War II had greater mobility, range, traverse, and elevation than their World War I predecessors, making them more powerful.[12]

As important as the new field guns were, the adoption of improved ammunition also had a significant impact. During World War I, shrapnel was the most prominent projectile employed by the combatants. Experience in the war, however, revealed that the projectile was of little value against a well-disciplined, trained army, especially when it was entrenched or behind fortifications. In light of this, most armies abandoned shrapnel in the mid–1930s for high-explosive steel shell, which had been around for years and fragmented into small splinters, to complement antitank, illuminating, smoke, and gas shells, and employed combination fuzes to detonate these projectiles.[13]

The long ranges of the fieldpieces that were appearing in the 1930s, the employment of camouflage, and the tendency to fire from protected defilade positions made ground observation to detect hostile batteries even more difficult than during World War I. These challenges led armies during the 1920s and 1930s to see aerial observation increasingly as a valuable method for finding targets behind enemy lines for their field guns to attack, because it could find targets that ground observers could not see.[14]

Although aerial observation had existed since before World War I to locate targets for artillery fire, the field artillery had to depend upon air forces for it, even though some officers had visualized the requirement for organic air observation for the field artillery. Because of their collective experience during and after the war, most European and American field artillerymen generally acknowledged that the air forces provided unsatisfactory aerial observation. Artillery commanders never knew when air observation would be available because they did not control it, while air force officers perceived that flying observation missions for the field artillery was not as high a priority as strategic bombing or other missions were and resisted flying observation missions for the field artillery.[15]

Dissatisfied with existing aerial observation procedures, European and American field artillerymen turned to organic aerial observation as a solution. Just prior to World War II, the British explored the feasibility of furnishing their field artillery with small, low-powered aircraft having specially trained personnel to spot for artillery fire, rather than using powerful fighter aircraft, which were awkward and moved too fast to locate targets. Ultimately, this would give the field artillery control over aerial observation, ensure the availability of aircraft, furnish trained spotters, and free the Royal Air Force from supplying air observation, a boring chore for most pilots. Ironically, resistance by the Royal Air Force for doctrinal reasons prevented the British army from actually implementing this type of air observation until 1941. The Royal Air Force insisted that field artillery officers could not be trained to the necessary standards required of pilots and that the aircraft would be shot down. Equally important, the Royal Air Force feared that the observation aircraft would be armed and that this would lead to an army air arm.[16]

As this was occurring, the Americans adopted organic air observation for their field artillery. Like the British, they too depended upon light aircraft and specially trained personnel for artillery spotting and reconnaissance and had to fight opposition from their air power people to implement organic air observation. Yet, as combat in World War II demonstrated, organic aerial observation proved to be invaluable. It permitted seeing targets behind enemy lines, troop concentrations, and other critical targets that could not be detected by ground observation and exploiting the long ranges of recently acquired guns and howitzers. Just as important, the presence of Allied organic air observers in the air over enemy territory often caused the Germans to cease firing and allowed Allied field artillery to fire with impunity.[17]

In the meantime, the Americans made a valuable breakthrough in fire control in their search for more effective observed indirect fire for close support. During World War I, they had two ways of massing fire from several batteries. One involved designating one or two points on a map as targets. This was satisfactory for scheduled fire but inadequate for close support because it was slow and rigid. The other method consisted of adjusting one battery on the target and then using it as a base for the others when maps were unavailable. As long as each battery adjusted successively and as long as each forward observer could see the target, aiming point, and battery, this method worked satisfactorily. However, when batteries adjusted simultaneously, forward observers had problems determining which battery was hitting the target because of dust. Likewise, massing fire was problematic when the forward observer could not see the target, aiming point, or battery because he had to see all three to adjust fire rapidly and effectively. In other words, shifting fire around the battlefield for close support was elusive because of the techniques of the time.[18]

With the appearance of more mobile motor vehicles, the Americans knew that furnishing close support by means of observed indirect fire with World War I techniques would be difficult.[19] Inspired by Lieutenant Colonel Neil Fraser-Tytler, a British field artillery officer in World War I who had demonstrated that artillery fire could be shifted from point to point on the battlefield, the American army under the leadership of Majors Carlos Brewer (q.v.) and Orlando Ward (q.v.) streamlined fire direction procedures in the 1930s and created the fire direction center, which centralized command, control, and communications by tying a battalion's firing batteries into one network. Employing the new methods and taking advantage of the fire direction center, a battalion's firing batteries could mass fire on a target in about twelve minutes and shift it around the battlefield as required. By 1941, employing the new fire direction techniques and the fire direction center, American field artillery could mass a division artillery of four battalions within five minutes after a call for fire, deliver a hammer blow without warning, and move barrages around as the battle shifted.[20]

Like the Americans, the Europeans reached the conclusion that World War I methods for massing fire were too slow and rigid for effective close support on a battlefield that was becoming increasingly more mobile. By 1941 the British, for example, had centralized command, control, and communications to permit delivering fire rapidly and accurately from division artillery and to engage a target of opportunity, which meant they were no longer tied to an uncompromising system of fire and had unprecedented flexibility.[21]

The appearance of reliable radio during the 1930s enhanced the new command, control, and communications networks and made observed indirect fire even more effective. Equipped with a radio, an observer could move forward with the other combat arms and quickly report changing conditions to the guns. This development severed the observers from the constraints imposed by telephone lines and gave them freedom to follow the ebb and flow of the battle freely. Together with the improved command, control, and communications organizations like the American fire direction center, radio-equipped air and ground observers fostered greater flexibility, enabling the field artillery to shift fire rapidly and effectively around the battlefield and making observed indirect fire a reality.[22]

Notwithstanding the introduction of new technology and techniques for fire direction, field artillery tactics remained fairly static during the 1920s and 1930s. Providing close support for the other combat arms still ruled, because World War I had reaffirmed its importance. Even though tanks had the potential of being a threat to the guns in future wars, most field artillerymen avoided developing antitank tactics. Because indirect fire had little effect upon tanks, field artillerymen knew that they would have to engage them with direct fire, which would make their field guns vulnerable to fire from tanks. When armies realized that using field artillery in antitank roles robbed it of its primary mission of furnishing close support, they developed towed artillery specially designed for antitank work in defensive situations. Some armies even introduced self-propelled antitank artillery to complement towed antitank guns and to function offensively.[23]

Although most armies placed antitank guns under the infantry, some armies created separate commands for self-propelled antitank artillery. The American army, for example, organized tank destroyer units composed of self-propelled antitank guns and developed aggressive doctrine that made them tank hunters. The development of antitank artillery, however, did not relieve the field artillery of the task of serving as antitank weapons if the situation warranted it, nor did it prevent self-propelled antitank artillery from serving as field artillery to deliver indirect fire if the situation justified such action.[24]

Likewise, even though armies saw the requirement for field artillery to support tanks on the offensive, they did not change their basic view about tanks. For the most part, they still viewed the tank as an infantry support weapon that moved at the pace of the infantry. Despite the rise of armored warfare advocates in Europe during the 1920s and 1930s, such as J.F.C. Fuller, B. H. Liddell Hart, and others, only the Soviets and Germans seriously combined field artillery, armor, and aircraft into a potent striking force. The Soviets declared that field artillery would clear the way for the tanks, centralized command of their artillery to provide massed fire, and devoted much of their resources to building a huge artillery force. In fact, by 1937 the Soviet army reputedly had over nine thousand field guns, roughly twice the number that the Germans had and three times the number that the French had. Yet, after the purges of the late 1930s had decimated the officer corps, the Soviets fell behind the Germans in developing armored warfare doctrine. This left the Germans as the only true proponent of armored warfare on the eve of World War II. Unlike the Soviets, the Germans decentralized command of their field artillery to support mobile armored columns.[25]

Armies also adopted antiaircraft artillery to engage aircraft so that their field artillery would not have to assume another mission that broke with tradition, except in an emergency. Although most armies equipped their divisions with antiaircraft artillery units under their own commanders to protect the division from aircraft attacks, some, such as the British, also formed separate commands, especially for defense of cities. As with field artillery, which frequently performed antitank missions, antiaircraft artillery often was pressed into service as an antitank weapon and field artillery piece. For example, the Germans employed their 88mm antiaircraft gun during the Spanish Civil War as a fieldpiece and antitank gun because of the lack of air targets, and later in North Africa and Russia, where the weapon achieved fame as an antitank gun.[26]

By World War II, antiaircraft and antitank artillery were serving as field artillery. Ideally, each should have performed that mission for which it had been designed, but combat and the need for firepower often dictated using each in a different role and at times blurred the distinctions among the three artilleries.

During World War II, the combatants further improved their field artillery. Of all the technological innovations that appeared between 1919 and 1945, the variable time fuze, called simply the VT fuze or proximity fuze, most truly revolutionized field artillery ordnance. Building upon work done by the British, American researchers tested the VT fuze in 1941. The fuze contained a miniature wet-cell battery and a five-tube, two-way radio. In contrast to conventional fuzes, which were preset before firing, the VT fuze detonated the round at a given distance from the target through reflected radio waves. The Americans first used the VT fuze in combat in antiaircraft artillery work in the Pacific theater, where duds could fall safely into the Pacific Ocean so that the enemy could not recover them and possibly duplicate them. After the fuze had demonstrated its dependability against the Japanese, the Americans adopted it for the field artillery.[27]

On 17 December 1944, during the Battle of the Bulge, the Americans employed the VT fuze for the first time against ground targets. Each shell burst above the ground at the precise height desired where its fragments could do the most damage against the Germans. The "funny fuze," as General George S. Patton called it, increased the effectiveness of American field artillery by as much as fifty times, as it had already done with antiaircraft artillery fire in the Pacific theater, and was one of the most critical but frequently overlooked technological developments of the war.[28]

Although rockets had fallen out of favor with the development of accurate rifled field artillery during the mid-nineteenth century, Germany, the Soviet Union, Great Britain, and the United States experimented with them during the 1930s and early 1940s to increase firepower and eventually adopted them during the war. The Soviets produced a system of truck-mounted multiple rocket launchers of several calibers from 75mm to 16 inches, named Katyusha, that fired rockets up to 6,000 yards. On the Leningrad front in September 1941, the Soviets employed their Katyusha rocket systems for the first time when they blasted German soldiers. During the latter months of 1942, the Soviets used them again effectively at the Battle of Stalingrad. In 1941, the Germans introduced their Nebelwerfer, a multiple rocket launcher that fired rockets up to 7,000 yards. As this was occurring, the Americans standardized a multiple rocket launcher late in 1944 that fired rockets up to 5,000 yards and used it for the first time during the Battle of the Hurtgen Forest in the fall of 1944. Although rocket barrages were spectacular and furnished tremendous amounts of firepower in a short period of time, they caused little damage to well-constructed fortifications.[29]

Rocket technology of the times had two major weaknesses. Outside of the German V–2 rocket, which had a range of about 200 miles, rockets were inaccurate for the most part and were employed to attack large-area targets, such as concentrations of personnel and equipment. Equally important, rockets gave off a heavy back-blast and left long smoke trails that made concealing them difficult and susceptible to counterbattery fire. Notwithstanding these problems, World War II rockets laid the groundwork for future developments and yielded large amounts of firepower rapidly when it was needed.[30]

As World War II aptly demonstrated, developments during the 1920s and 1930s eliminated many of the deficiencies that had plagued the field artillery during World War I. Motor vehicles improved mobility, while the new field guns had ranges and firepower that were superior to those of their predecessors. Meanwhile, self-propelled artillery furnished mobility and firepower and combined with tanks to furnish unprecedented striking power for armored formations. As critical as motorization and new weapons were, the appearance of new fire control techniques and portable field radios provided the most striking difference between World War II and World War I field artillery capabilities. Teamed with air or ground observers equipped with radios, fire control systems allowed gun crews to shift observed indirect fire rapidly and effectively around the battlefield to engage targets, and they produced unparalleled flexibility. As a result, the days of rigid firing plans, characteristic of World War I, disappeared for most armies during World War II, especially the American. Observed indirect fire that could furnish effective close support for a mobile army had become a reality and disarmed its critics.

ADAPTING TO THE NUCLEAR AGE

Following World War II, the Cold War between the United States and the Soviet Union generated an arms race that included the introduction of tactical nuclear artillery—rockets, guided missiles, and tube artillery—and lasted through the 1980s. Since firepower was one of the primary lessons of World War II, the Americans, who had become a leader in field artillery developments during the war, saw free-flight rockets and guided missiles as possible weapons to upgrade the firepower and range of their field artillery.[31] Although different forms of guided missiles existed (surface-to-surface, surface-to-air, air-to-surface, and air-to-air), surface-to-surface missiles furnished the field artillery with the greatest potential of reaching targets beyond the range of existing tube artillery and rockets and, therefore, attracted the most attention from combat developers. Full-scale work to develop rockets and missiles began in 1947, when American and captured German scientists tested American-adapted V–2 rockets to acquire critical information on the upper atmosphere and collect data on rocketry. Using knowledge gathered from experimental rockets and test flights during the last years of the 1940s, the American army adopted the Corporal and Redstone guided missiles with ranges of 80 and 200 miles, respectively, during the mid–1950s. These missiles carried conventional and nuclear warheads for employment against troop concentrations, marshaling yards, supply depots, vital command centers, and beachheads.[32]

During the decade after 1945, the Americans also produced free-flight rockets that lacked guidance systems, such as the Honest John, fielded in the mid–1950s, with a range of 24 miles. The Honest John and other free-flight rockets were aimed and fired from their launchers like a cannon and could carry a conventional or nuclear warhead. Because American guided missiles and rockets were the leading edge of such technology in the 1950s, many Western European countries adopted them rather than developing their own because of the expense of producing such weapons.[33]

Although the free-flight rockets and missiles of the first generation were inaccurate, they still gave armies artillery weapons with tremendous destructive power. In fact, a single 20-kiloton warhead, such as that on the Honest John or Corporal, had the equivalent power of four hundred thousand conventional 155mm projectiles. Anticipating the future of guided missiles and free-flight rockets, some Soviet and American army officers of the 1950s confidently proclaimed that such weapons would eventually replace tube artillery.[34]

The real breakthrough in artillery firepower came before the advent of guided missiles and free-flight rockets that could carry nuclear warheads. As they were working on these forms of artillery, the Americans introduced a 280mm artillery piece that could fire a nuclear projectile. Known as Atomic Annie, the gun fired the first atomic artillery shell on 25 May 1953 in a demonstration at Frenchman's Flat, Nevada. For the field artillery this opened up a totally new era in firepower by demonstrating that the technology existed to make atomic projectiles for field cannon and provided the foundation for the tactical nuclear battlefield. Aware that the 280mm gun was heavy and difficult to maneuver, the Americans developed a nuclear projectile for their 8-inch howitzer late in the 1950s and one for their 155mm howitzer early in the 1960s. This combination of guided missiles, free-flight rockets, and tube artillery with nuclear warheads gave the Americans unsurpassed firepower and a means of offsetting the Soviet Union's numerical superiority in Europe and massive quantities of firepower. Although the weapons did not change the field artillery's traditional mission of supporting the other combat arms, tactical nuclear field artillery did create the potential for unimaginable destruction.[35]

Because firepower and mobility were two of the primary lessons of World War II and because tactical nuclear warfare demanded more dispersion and even more mobility and protection for ground forces than earlier warfare had, interest in obtaining new, improved self-propelled artillery heightened early in the 1950s. Dissatisfied with their World War II self-propelled pieces, the Americans designed a new family of self-propelled artillery shortly after the war with more mobility, firepower, and traverse. During the Korean War (1950–1953), the Americans learned, much to their dismay, that postwar designs were hopelessly obsolete because they lacked 360-degree traverse, high-angle fire capabilities, and sufficient mobility and lethality to fight an enemy that used encircling tactics and to engage targets on the reverse side of steep mountain slopes. The emerging tactical nuclear battlefield made these characteristics even more critical for the field artillery to furnish effective close support and to survive, and concurrently highlighted the necessity of providing an enclosed shelter on self-propelled artillery to protect the crew from radiation as well as other battlefield hazards.[36]

In light of these deficiencies identified early during the Korean War, the Americans fielded a new family of self-propelled artillery in the 1960s. These new self-propelled howitzers had 360-degree traverse and 75-degree elevation. To protect the gun crew from small arms fire, radiation, and other hazards, they had an enclosed crew shelter. Manufacturers stripped these pieces of cumbersome external recoil mechanisms and breech assemblies designed for operation in the close quarters of an armored vehicle, and mounted the tube on a carriage specially developed for artillery. This was an important break with previous American self-propelled artillery, because these weapons had consisted of standard towed artillery tubes that had been lifted off their carriages and installed on modified tank hulls to give them tanklike characteristics. Because American self-propelled artillery of the early 1960s represented the vanguard of self-propelled artillery technology, many Western European countries adopted it for their armies in the 1960s to supplant older American self-propelled artillery that had been the mainstays in their armies since World War II, while other countries throughout the world obtained obsolete American towed and self-propelled artillery for their armies.[37]

Some countries eventually started manufacturing their own self-propelled artillery with 360-degree traverse, high-angle fire capabilities, and crew shelters. The British, for example, produced their own self-propelled weapons for fighting on the nuclear battlefield, rather than depending upon the Americans. Thus, as the 1960s were opening, a totally new family of field artillery was appearing in the United States and Western Europe that differed significantly from its ancestors of the 1940s and 1950s. These new self-propelled pieces had the ability of supporting the other combat arms on mobile conventional and nuclear battlefields with effective indirect fire.[38]

To keep abreast of American and European field artillery developments during the early years of the Cold War and to replace worn-out World War II pieces, the Soviets standardized new field artillery pieces during the late 1950s. They added a new 203mm field gun with a range of 29,000 meters (31,900 yards) that fired conventional and nuclear rounds. They also adopted a new 152mm field howitzer having a maximum range of 17,300 meters (19,000 yards) with high-explosive shell. Both the 203mm howitzer and the 152mm howitzer were also employed by the Soviets' Eastern European allies. Together, these two pieces gave Soviet and Warsaw Pact field artillery more mobility and lethality than previous weapons had offered and could be rapidly emplaced and dispersed on the nuclear battlefield. In addition to these new towed weapons, the Soviets adopted new self-propelled artillery. In designing these towed and self-propelled fieldpieces and others, the Soviets focused upon mobility, firepower, simplicity, and standardization to close the gap between American and their own field artillery, and included turrets to protect the crew.[39]

Simultaneously with the push for more powerful field artillery, the drive to support the other combat arms that were becoming increasingly more mobile and lethal led to the use of aircraft and helicopters as prime movers. In World War II combatants introduced parachute troops to take advantage of the superior mobility of aircraft to ground vehicles. Using aircraft, they lifted troops over obstacles, such as enemy troops, and dropped them on critical objectives. Although aircraft gave armies tremendous mobility, the troops were vulnerable to heavily armed forces. In light of this, the Americans adopted a 75mm pack howitzer that could be airlifted and dropped by parachute to increase the firepower of airborne units. By the late 1950s, the improved carrying power of aircraft and the development of air transportable field artillery permitted the American army to airlift towed 105mm and 155mm howitzers as well as Honest John and Little John rockets more easily than their predecessors of World War II. For the Americans air transportable field artillery was critical because they required a method of transporting field artillery into Europe rapidly if war broke out there.[40]

Although the American army was working on methods of transporting field guns by air during the late 1950s and early 1960s in response to the need for more mobility on the tactical nuclear battlefield than towed and self-propelled artillery offerred, the development in the early 1960s of aerial rocket artillery, a helicopter with a rocket system attached, was an even more important breakthrough. Unrestrained by roads and terrain, aerial artillery had the ability to fly along the nap of the land, to fire its rockets at the target, and then to fly off to engage more targets. Moreover, this type of artillery could fire from the air or ground, and furnish direct or indirect fire as required.[41]

Ironically, it was the Vietnam War, dominated by small-unit action, rather than tactical nuclear warfare, that vindicated the supporters of aerial rocket artillery and helicopter-transported, towed 105mm and 155mm howitzers. Repeatedly throughout the war, American field artillerymen rapidly moved their howitzers by helicopter and flew aerial rocket artillery in support of the ground forces on terrain that prohibited moving towed artillery into positions with their normal prime movers—trucks. Taking advantage of the mobility of helicopters, the Americans even conducted artillery raids from fire bases into remote places by displacing artillery to supplementary positions to attack targets beyond the ranges of artillery at the main fire base. Field artillerymen would quickly fly their weapons into position, fire the required number of missions, and then rapidly displace to avoid enemy counterattacks.[42]

At the same time, the Vietnam War forced the field artillery to make radical tactical changes to fight a war that lacked fronts along the lines of World War II. Because American field artillery could provide fast, accurate fires to protect the other combat arms, the enemy made the guns prime targets for infiltration or full-scale attacks. Unable to perform their missions and simultaneously protect themselves, the Americans built fire support bases that were often mutually supporting by being within range of each other. The Americans positioned their guns with the command post of a maneuver unit. From the base, maneuver commanders conducted offensive operations under the protective umbrella of the field artillery. Such an arrangement provided fast, effective fire support and reinforced the growing American reliance upon firepower rather than maneuver for defeating the elusive enemy.[43]

As post–World War II developments suggested, the United States and the Soviet Union and their respective allies modernized their field artillery to fight on a tactical nuclear battlefield. With improvements in technology, scientists could reduce the size of a nuclear warhead to fit on an 8-inch, a 152mm, or 155mm projectile by the beginning of the 1960s and had even developed rockets and guided missiles with the capabilities of carrying a tactical nuclear warhead. At the same time, armies developed tactics and doctrine that focused on mobility and dispersion to negate the power of nuclear weapons. The combination of tactics, doctrine, mobility, and new technology opened new vistas for the field artillery by the mid–1960s that provided unsurpassed mobility and firepower.

HIGH TECHNOLOGY AND THE FIELD ARTILLERY

The Arab-Israeli War of October 1973 portended future warfare on the high-technology battlefield. During the war, remotely piloted vehicles (RPVs) gave the field artillery over-the-hill and near real-time target acquisition capabilities, while first-generation precision-guided munitions provided unsurpassed accuracy and reaffirmed their value, as they had already done during the Vietnam War. Likewise, multiple rocket launchers confirmed their ability to neutralize air defense sites and other large static targets with tremendous firepower. These and other weapons significantly increased the lethality of combat and added an entirely new dimension to the field artillery.[44]

Impressed with the firepower of modern weapons as exhibited by the Arab-Israeli War and confronted with numerically superior Soviet and Warsaw Pact (an alliance involving the Soviet Union and Eastern European countries for mutual protection) armed forces that had been modernized and enlarged during the 1960s and early 1970s, the Americans and their North Atlantic Treaty Organization (a military alliance involving the United States and Europe) allies launched a massive rearmament effort in the 1970s and 1980s to incorporate high-technology weapons to offset their opponent's imposing military forces.[45]

With improved assembly, design, and aerodynamics, all of which improved accuracy and range over their World War II ancestors, multiple rocket launchers attracted attention after being overshadowed by more glamorous free-flight rockets and guided missiles in the 1950s and 1960s. Once armies had become convinced of the tremendous firepower that these rocket systems had, they started building their own, even though they were expensive. The West Germans introduced their Light Artillery Rocket System (LARS) in 1969 while other countries were still contemplating the value of multiple rocket launchers. Early in the 1980s, the Americans adopted a multiple rocket launcher, called the Multiple-Launch Rocket System (MLRS). By the middle of the 1980s, France and even non-European countries, such as Brazil, were equipping their armies with multiple rocket launchers in their quest for more firepower.[46]

In response to the Soviet introduction of improved guided missiles in the 1970s, the Americans introduced the Pershing I, Pershing IA, and Pershing IB, as multiple rocket systems were being developed. The Pershing II, which was fielded in the 1980s, represented an evolutionary improvement over its predecessor, had the ability of delivering a nuclear warhead up to 1,800 kilometers (1,125 miles), and could strike a target in Soviet soil six minutes after being launched in West Germany. These new guided missiles fielded by the Soviet Union and the United States raised the specter of nuclear holocaust even more and led to various attempts to abolish them through treaties. The most successful treaty, the Intermediate-Range Nuclear Forces Treaty, signed late in 1987, outlined provisions to eliminate the Pershing II, Pershing IA, Pershing IB, and Ground-Launched Cruise Missile, and the Soviet SS–4, SS–5, SSCX–4, SS–12, SS–20, and SS–23 missiles to reduce the nuclear threat in Europe.[47]

Convinced of the need to attack second-echelon Soviet and Warsaw Pact forces in defense of Western Europe, which the Pershing missiles did not address, the Americans obtained the Army Tactical Missile System during the latter years of the 1980s to replace the Lance missile, which had been fielded in the 1960s. Fired by the Multiple-Launch Rocket System, the Army Tactical Missile System, with a range of over 100 kilometers, dispensed submunitions designed to neutralize lightly armored vehicles over the target area.[48]

Despite improvements in accuracy, range, and lethality during the late 1960s and 1970s as indicated by the new technology, tube artillery was falling behind developments in armor as guided missile technology was becoming more sophisticated. Even newer self-propelled artillery lacked the mobility of the new tanks being fielded in the 1980s. The field artillery was losing the ability to supply effective observed indirect fire to the other combat arms because it could not keep up with the faster tanks and infantry fighting vehicles. The American self-propelled M109 155mm howitzer, which was one of the best self-propelled artillery weapons in the world, was slow, had a relatively slow rate of fire, and lacked an automated ammunition system, an armored ammunition vehicle, and sufficient armor. Because of this, the United Kingdom, West Germany, and Italy launched an effort in 1973 to replace the M109 as the self-propelled 155mm howitzer standard in Europe with the SP70 self-propelled 155mm howitzer. As planned, the SP70 would be the self-propelled howitzer of the future, more lethal, reliable, and effective than the M109, having a range of 24 kilometers (30 kilometers with rocket-assisted projectiles), and having the mobility of the modern battle tank and infantry fighting vehicles that were scheduled for adoption in the 1980s. The SP70 was vastly superior to the weapons that it was to replace. However, production delays eventually led to its abandonment in 1987 and forced West Germany, Italy, and the United Kingdom to start their own separate 155mm self-propelled howitzer development projects.[49]

The Americans decided that the SP70 would not meet their need to supplant the M109 and opted to develop their own divisional support weapon system. Early in the 1980s, they devised the Howitzer Extended Life Program and later the Howitzer Improvement Program to upgrade the M109 but simultaneously realized that enhancements to the weapon were only interim measures. The Americans really required a totally new self-propelled 155mm howitzer. With this in mind, in the mid–1980s they planned to field the Advanced Field Artillery System sometime late in the 1990s or early in the 2000s. This would be an entirely new fieldpiece and not an enhanced version of an older model; it would take advantage of emerging technology to improve dramatically firepower, survivability, mobility, and accuracy.[50]

The United States and its European allies also required first-round accuracy (hitting the target with the first round without registration) that new field artillery weapons systems alone could not furnish. Such capabilities would help offset the numerical superiority of Soviet and Warsaw Pact armies, especially their armored forces, by reducing the number of rounds required to neutralize or destroy a target. Unfortunately, the state of artillery munitions limited accuracy and prevented achieving first-round accuracy. Into the 1970s the American army as well as other armies still relied upon various kinds of unguided, high-explosive munitions. The most common, because it was the most useful and had the greatest variety of applications, was a high-explosive shell that consisted of a hollow steel case filled with one of a number of chemical explosives to engage personnel and materiel. In the conventional indirect fire role, the most common shell after high-explosive was white phosphorous. When detonated, it created smoke to provide cover or mark a target. The field artillery also had illuminating shells, various kinds of antitank rounds, such as armor-piercing and high-velocity armor-piercing that penetrated armor by kinetic energy, and high-explosive antitank.[51]

In the search for more accurate and lethal projectiles, many armies introduced improved conventional munitions. Development of improved conventional munitions began in the United States late in the 1960s, with other countries following suit. Essentially, improved conventional munitions consisted of a carrier shell loaded with submunitions. The carrier shell delivered the submunitions over the target, where it was detonated to liberate the submunitions to furnish the tactical effect. The centrifugal force caused by the spinning carrier shell distributed the submunitions in a more or less regular pattern to give better accuracy than existing conventional munitions and more lethality.[52]

Against tanks and other armored vehicles, such as self-propelled artillery, most artillery projectiles, including improved conventional munitions, were ineffective because of the number of rounds required to destroy a tank or armored vehicle. During World War I, World War II, the Korean War, and the Vietnam War, studies conducted by the American army indicated that about 1 percent of the tank kills came from field artillery. In some situations, as many as 1,500 high-explosive rounds were required. The same studies projected that about 300 improved conventional munition rounds were needed, and firing 300 to 1,500 rounds demanded extensive logistical support, was expensive and ineffective, and wore out cannon tubes rapidly.[53]

Because existing munitions lacked sufficient accuracy to kill moving armored targets and because the Soviets and their Eastern European allies were numerically superior in armored (field artillery and tanks) vehicles to the United States and its Western European allies, the American army led the way to balance the ratio.[54] Impressed with semi-smart munitions employed in the Arab-Israeli War of 1973 and the latter years of the Vietnam War, the Americans introduced the Copperhead 155mm cannon-launched guided projectile late in the 1970s after several years of work. The projectile depended upon a laser designator operated by a ground or air observer to guide it to the target. At ranges of up to 16 kilometers, American field artillery could achieve direct hits with the Copperhead. This meant firing fewer rounds to hit a target and less wear on the artillery tube. In light of this, the Americans concluded that their field artillery would have the capability of destroying large numbers of tanks before they reached the main battlefield and simultaneously neutralizing enemy artillery that was supplying suppressive fire on friendly antitank guided weapons. In other words, Copperhead would help offset the numerical disparity between the Warsaw Pact and North Atlantic Treaty Organization armies and provide first-round accuracy.[55]

The search for the technology to neutralize the Soviet and Warsaw Pact numerical superiority continued to encourage the Americans to find projectiles that were even more effective than semi-smart munitions such as the Copperhead and led to the development of smart munitions (terminally guided) or "fire-and-forget" munitions in the 1980s. Unlike the first generation of precision-guided munitions, which depended upon a laser beam to direct the projectile to the target, a "fire-and-forget" munition had electronic onboard seekers and microprocessors to enable it to home in on a recognizable target without employing a laser designator. During the 1970s and 1980s, the Americans worked on a Sense-and-Destroy Armor Munition (SADARM) for their 155mm howitzer and Multiple-Launch Rocket System and a Terminally Guided Warhead (TGW) for their Multiple-Launch Rocket System, and anticipated having them fielded sometime early in the 1990s to add a new level of lethality to the battlefield.[56]

As envisioned, SADARM would be a shell that carried submunitions. Upon reaching the target area, the carrier munition would eject the submunitions. Each submunition would have a sensor and deploy a parachute designed to cause the submunition to rotate as it descended. When a target was detected by the sensor, a metal slug would be fired from the submunition toward the target. The Americans planned to employ SADARM and other terminally guided munitions to attack deep targets, offset the Warsaw Pact's numerical and firepower superiority, and effectively neutralize tanks and self-propelled field artillery. Because of the accuracy of "fire-and-forget" projectiles, the field artillery would have the ability to hit targets with fewer rounds and would even have the capability of hitting a target with the first round. Yet, such munitions were expensive and could be used only against high-value targets.[57]

To exploit the new munitions and weapon systems with their unsurpassed ranges and accuracy demanded equally sophisticated target acquisition systems to detect targets, provide real-time intelligence, and lase targets for precision-guided munitions. During the closing weeks of World War II, the British army began experimenting with the use of radar to detect mortar bombs and determine their point of origin. Although the system was cumbersome, it worked and was gradually improved upon during the following years. With the electronic revolution that followed the invention of the transistor, computer-based radar became available to locate a mortar battery and even an artillery battery that was even more difficult to find. Unlike the mortar, which had a consistent parabolic trajectory, field artillery had a multiplicity of trajectories that were based upon the projectile employed, powder charge, and kind of artillery (gun or howitzer).[58]

The first effective computer-based artillery and mortar-locating radars appeared in the 1960s. These were soon replaced by the British Cymbeline and the American AN/TPQ–36 Mortar Locating Radar and the AN/TPQ–37 Weapon Locating Radar late in the 1970s. Using a radar beam, the radars scanned the horizon, tracked anything that they saw, and computed the origin of the projectile before it could even hit the ground. The introduction of radars in the 1970s and 1980s led the American, European, and several Middle and Far Eastern armies to discard sound-and-flash ranging because the latter was obsolete.[59]

Perhaps the most glamorous target acquisition system was the unmanned aerial vehicle. With the development of complex ground air defenses in the 1960s and early 1970s, reconnaissance, target acquisition, and adjusting fire by employing fixed-wing aircraft or helicopters became more precarious, as the Arab-Israeli War of 1973 testified. In view of this, various armies, such as the Israeli and American, seriously explored the possibilities of performing aerial observation with unmanned aerial vehicles that were equipped with a television camera, conventional cameras, and other types of sensors as substitutes for manned aircraft. Although costs and technical problems stalled the American army's effort to develop the Aquila Remotely Piloted Vehicle late in the 1980s and prompted it to begin work on a relatively inexpensive family of unmanned aerial vehicles for the 1990s, the Israeli unmanned aircraft was fielded, as was the British remotely piloted vehicle. In the meantime, although most European remotely piloted vehicle efforts suffered setbacks during the 1970s because of the costs and the difficulty of producing such technology, European countries still outlined plans in the 1980s to introduce remotely piloted vehicles to give their armies over-the-hill target acquisition capabilities by the 1990s that ground observers could not furnish.[60]

Wishing to match Soviet and Warsaw Pact firepower, the Americans and Europeans also turned to improved fire control systems. After years of design and development dating back to the late 1940s of various kinds of computers, the British army introduced its first computer for calculating fire direction data in 1961. Within years the Americans adopted the Field Artillery Digital Automatic Computer (FADAC) for the tactical nuclear battlefield. Even though the computer did not totally eliminate the need for manual computation of firing data, it greatly improved the field artillery's performance by reducing fatigue and errors by fire direction personnel and by permitting surprise fires. By doing this, the Field Artillery Digital Automatic Computer greatly enhanced accuracy, decreased response time by calculating target information and providing almost instantaneous firing data, and permitted gun crews to fire longer missions and hit more targets with less ammunition.[61]

Shortly afterward, other armies fielded computers for calculating fire direction information. In 1969 the British army introduced Field Artillery Computer Equipment (FACE) into service. Besides decreasing response time and improving accuracy, the computer reduced training time, replaced slide rules and firing tables, and took into account meteorological data. In 1973–1974 the Swedish army adopted a computer for calculating fire data. About the same time, the West Germans introduced a computer into its army for calculating firing data, as did the Norwegians.[62]

In the 1970s second-generation computers began appearing. During that decade, the Americans adopted the Tactical Fire Direction System (TACFIRE). In 1978 the British announced their intention of developing the Battlefield Artillery Target Engagement System (BATES), with fielding sometime in the 1990s. In 1980 the French placed ATILA into service. It was a fully integrated data transmission and computer system that linked the forward observer with the fire control computer at the regimental headquarters that made the fire mission decisions. Because of these computers and others that were being introduced in the 1970s and 1980s, the field artillery could mass fire more rapidly and accurately than ever before. Simultaneously, computers made possible achieving first-round accuracy by improving the precision of fire direction data.[63]

With the increased tempo of battle, the appearance of more mobile armored vehicles, and the requirement for better command, control, and communications, armies searched for better ways to tie their target acquisition systems and weapons into an effective fire support network than existing fire control computers permitted. Early in the 1980s, the Americans decided to replace their Tactical Fire Direction System, which was 1960s technology, with the Advanced Field Artillery Tactical Data System in the 1990s, which would offer distributive processing rather than centralized processing as the previous system did. For the Americans and their North Atlantic Treaty Organization allies better command, control, and communications systems loomed particularly important because they had to upgrade their field artillery productivity to offset the numerical advantage of Soviet and Warsaw Pact fire support systems. As a result, the West assumed the lead in the 1980s in developing sophisticated command, control, and communications systems to ensure effective fire support.[64]

The high-technology revolution of the 1970s and 1980s dramatically transformed the field artillery. With the appearance of computers for command, control, and communications, precision-guided munitions, radars and remotely piloted vehicles for target acquisition, and new weapon systems, the field artillery acquired unsurpassed lethality, mobility, and ability to acquire targets. Yet, this new technology was expensive and had not been proven in combat.[65]

Early in 1991, the Persian Gulf War between the United States together with its allies (Great Britain, France, Egypt, Saudi Arabia, and various Arab states) and Iraq provided the field artillery with the first opportunity to employ in combat the new high technology that had been introduced since the Arab-Israeli War of 1973. Armed with precision-guided munitions, other tremendously lethal munitions, and the new Army Tactical Missile System, Allied field artillery demonstrated unparalleled accuracy and could actually destroy a target with only one round if needed. One anecdote from the war described an Iraqi position that was attacked by massed dual-purpose improved conventional munition fire from American field artillery. The defenders scurried to a nearby bunker only to have a Copperhead round fly into the laser-designated door of the bunker. Realizing the futility of flight, the rest of the enemy unit surrendered. Multiple rocket launchers simultaneously silenced Iraqi artillery, permitting the maneuver arms to move with little fear of enemy artillery fire, and demonstrated their lethality. Army Tactical Missile Systems, for example, destroyed or rendered combat-ineffective every target that they attacked at ranges far beyond any existing artillery weapon. Likewise, sophisticated target acquisition radars identified Iraqi artillery and other fire support systems for effective counterbattery fire from Allied indirect fire systems, while British remotely piloted vehicles provided solid targeting information, reconnaissance data, and battle damage assessment in near–real time to aid the Allied fire support effort.[66]

Even so, the war reinforced the need for even better field artillery systems. Command, control, and communications systems, target acquisition systems, and weapon systems would require even more mobility in the future because the Persian Gulf War illustrated the tremendous mobility of maneuver forces and the difficulties that field artillery systems often had keeping up with them. Yet, by the beginning of 1992, participants and observers were only beginning to assess the lessons of the war by making tentative observations about what had to be done to improve their field artillery for the future.

However, the end of Cold War with the collapse of the Warsaw Pact in 1991 and breakup of the Soviet Union that same year eliminated the driving force behind fielding new and improved field artillery materiel, especially in Europe and the United States, which were among the leaders in introducing state-of-the-art materiel. Only time would tell if the materiel on the drawing boards at the beginning of the 1990s, introducing new levels of sophistication, would ever be adopted by the former Cold War adversaries.

As suggested, the introduction of expensive high technology in the form of computers for fire direction, target acquisition radars, remotely piloted vehicles, and precision-guided munitions during the 1960s, 1970s, and 1980s capped a revolution of seventy years. During those years, the field artillery abandoned the horse as a prime mover for towed and self-propelled weapons and adopted sophisticated fire direction control techniques, organic aerial observation, unmanned aerial vehicles, extremely effective fuzes, and nuclear projectiles. Altogether, these technologies permitted the field artillery to shift fire around the battlefield rapidly and accurately to support the other combat arms with effective observed indirect fire. The high-technology revolution of the latter decades of the twentieth century gave the field artillery an unsurpassed ability to move, shoot, and communicate—the dream of field artillerymen for centuries.

NOTES

1. Shelford Bidwell and Dominick Graham, *Firepower: British Army Weapons and Theories of War, 1904–1945* (London: George Allen and Unwin, 1982), 173; Christopher Bellamy, *Red God of War: Soviet Artillery and Rocket Forces* (London: Brassey's Defence Publishers, 1986), 154.

2. Frank E. Comparato, *Age of Great Guns: Cannon Kings and Cannoneers Who Forged the Firepower of Artillery* (Harrisburg, PA: The Stackpole Company, 1965), 223–24.

3. Bellamy, *Red God of War*, 156; H.C.B. Rogers, *A History of Artillery* (Secaucus, NJ: The Citadel Press, 1975), 170–71; John Batchelor and Ian V. Hogg, *Artillery* (New York: Charles Scribner's Sons, 1975), 170–71.

4. Bellamy, *Red God of War*, 156; Rogers, *A History of Artillery*, 170–71; Batchelor and Hogg, *Artillery*, 170–71.

5. Comparato, *Age of Great Guns*, 224–25.

6. Harry C. Thomson and Lido Mayo, *The Ordnance Department: Procurement and Supply* (Washington, DC: Office of the Chief of Military History, U.S. Army, 1960), 72; John P. Lucas, "The 105mm Howitzer," *Field Artillery Journal*, February 1941, 69.

7. Ian V. Hogg, *The Guns: 1939–45* (New York: Ballantine Books, 1970), 35–41; Batchelor and Hogg, *Artillery*, 22.

8. Andrew W. Hull, "Evolution of Soviet Self-Propelled Artillery," *Field Artillery Journal*, March-April 1978, 8–10; Bellamy, *Red God of War*, 45; Curt Johnson, *Artillery:*

The Big Guns Go to War (London: Octopus Books, 1975), 102–4; Batchelor and Hogg, *Artillery*, 114–22.

9. Rogers, *A History of Artillery*, 170–72; Bidwell and Graham, *Firepower*, 156, 158.

10. Comparato, *Age of Great Guns*, 161, 248–49, 352–53; Johnson, *Artillery*, 89; Jonathan B.A. Bailey, *Firepower and Field Artillery* (Oxford: The Military Press, 1989), 210–11.

11. Ian V. Hogg, *Artillery in Color* (New York: Arco Publishing, 1980), 12–16.

12. Bellamy, *Red God of War*, 46, 48; Robert M. Danford, Lecture, U.S. Army War College, 23 Sep 1939, 28–31, in Morris Swett Technical Library, Fort Sill, OK; Hogg, *Artillery in Color*, 16–17; Hogg, *The Guns*, 30–57; Johnson, *Artillery*, 72–74; The U.S. Army Field Artillery School, *History of the Development of Field Artillery Materiel* (Fort Sill, OK: The Field Artillery School, 1939), 75–76.

13. Hogg, *The Guns*, 17, 24; Batchelor and Hogg, *Artillery*, 23.

14. Boyd L. Dastrup, *King of Battle: A Branch History of the U.S. Army's Field Artillery* (Fort Monroe, VA: Office of the Command Historian, U.S. Army Training and Doctrine Command, 1992), 206; Kenneth Brookes, *Battle Thunder: The Story of Britain's Artillery* (Reading, England: Osprey Publishing Ltd., 1973), 187–88.

15. William W. Ford, "Grasshoppers," *U.S. Army Aviation Digest*, June 1982, 3–4; Bidwell and Graham, *Firepower*, 101–3, 262–63. Aerial observation had been around since the 1790s, when the first tethered balloons had been used to locate enemy forces. However, aerial observation for field artillery fire was impractical until the development of indirect fire late in the nineteenth century.

16. Bidwell and Graham, *Firepower*, 262–63.

17. Richard P. Weinert, Jr., *A History of Army Aviation: 1950–1962*, rev. ed. (Fort Monroe, VA: Office of the Command Historian, U.S. Army Training and Doctrine Command, 1991), 2–6; Shelford Bidwell, *Gunners at War: A Tactical Study of the Royal Artillery in the Twentieth Century* (Aberdeen, Scotland: Arms and Armour Press, 1970), 103–17; Bidwell and Graham, *Firepower*, 262–63; Kenneth Brookes, *Battle Thunder*, 187–89; Dastrup, *King of Battle*, 220.

18. Frank G. Ratliff, "The Field Artillery Fire Direction Center—Its Past, Present, and Future," *Field Artillery Journal*, May-June 1950, 117; Comparato, *Age of Great Guns*, 241–42.

19. Comparato, *Age of Great Guns*, 241.

20. Russell Gugeler, "Fort Sill and the Golden Age of Field Artillery," 8–9, unpublished manuscript in Morris Swett Technical Library, Fort Sill, OK; Ratliff, "The Field Artillery Fire Direction Center," 118; Comparato, *Age of Great Guns*, 241–45.

21. Graham and Bidwell, *Firepower*, 162–65, 198–201, 211–12, 215; Bidwell, *Gunners at War*, 130–44; Comparato, *Age of Great Guns*, 242; "Field Artillery Notes," *Field Artillery Journal*, January–February 1932, 111.

22. Bailey, *Field Artillery and Firepower*, 162–65.

23. Bellamy, *Red God of Fire*, 187; Batchelor and Hogg, *Artillery*, 82; Dastrup, *King of Battle*, 214.

24. Bellamy, *Red God of War*, 187; Batchelor and Hogg, *Artillery*, 82; Dastrup, *King of Battle*, 214.

25. Jonathan M. House, *Towards Combined Arms Warfare: A Survey of Tactics, Doctrine, and Organization in the 20th Century* (Fort Leavenworth, KS: Combat Studies

Institute, U.S. Army Command and General Staff College, 1984), 43–78; Bailey, *Field Artillery and Firepower*, 156–58; Bellamy, *Red God of War*, 45–48.

26. Batchelor and Hogg, *Artillery*, 69, 84, 101; Bailey, *Field Artillery and Firepower*, 159–61; Brookes, *Battle Thunder*, 163, 201–13; Dastrup, *King of Battle*, 208; Comparato, *Age of Great Guns*, 137; Johnson, *Artillery*, 87.

27. Johnson, *Artillery*, 109; Comparato, *Age of Great Guns*, 265–67; Bernard and Fawn Brodie, *From Crossbow to H-Bomb* (Bloomington, IN: Indiana University Press, 1973), 213–14.

28. Johnson, *Artillery*, 109; Bellamy, *Red God of War*, 201; Comparato, *Age of Great Guns*, 267; Brodie, *From Crossbow to H-Bomb*, 213–14.

29. Comparato, *Age of Great Guns*, 288–304; Brodie, *From Crossbow to H-Bomb*, 226–32.

30. Comparato, *Age of Great Guns*, 288–304; Johnson, *Artillery*, 91.

31. Nels A. Parson, Jr., *Guided Missiles in War and Peace* (Cambridge, MA: Harvard University Press, 1956), 10.

32. Johnson, *Artillery*, 130–31; Parsons, *Guided Missiles in War and Peace*, 124–26; A. R. Weyl, *Guided Missiles* (London: Temple Press Ltd., 1949), 77–80; Werner von Braun and Frederick I. Ordway III, *History of Rocketry and Space Travel* (New York: Thomas Y. Crowell Company, 1966), 123–26; Holger N. Toftoy, "Army Missile Development," *Army Information Digest*, December 1956, 10–32.

33. Rogers, *A History of Artillery*, 211; Brookes, *Battle Thunder*, 235–37; "The Soviet Union's Missile Artillery," *Artillery Trends*, December 1959, 24–27; Toftoy, "Army Missile Development," 32; Bellamy, *Red God of War*, 155–56.

34. Comparato, *Age of Great Guns*, 298–300; Johnson, *Artillery*, 130–31.

35. Brookes, *Battle Thunder*, 235; Johnson, *Artillery*, 119.

36. Philip H. Stevens, *Artillery through the Ages* (New York: Franklin Watts, Inc., 1965), p. 176; Comparato *Age of Great Guns*, 308–9; Brookes, *Battle Thunder*, 234; Hogg, *Artillery in Color*, 171.

37. Stevens, *Artillery through the Ages*, 176; Comparato, *Age of Great Guns*, 308–9; Brookes, *Battle Thunder*, 234; Hogg, *Artillery in Color*, 171.

38. Shelford Bidwell, ed., *Brassey's Artillery of the World* (New York: Bonanza Books, 1979), 72.

39. Christopher F. Foss, *Artillery of the World*, rev. ed. (New York: Charles Scribner's Sons, 1976), 55–56; Andrew W. Hull, "Evolution of Soviet Self-propelled Artillery," *Field Artillery Journal*, March-April 1978, 9–10; Johnson, *Artillery*, 128; Ian V. Hogg, *Artillery 2000* (London: Arms and Armour Press, 1990), 73.

40. Stevens, *Artillery through the Ages*, 177.

41. U.S. Army Artillery and Missile School, "Field Artillery Weapons, 1964"; "Field Artillery Weapons," *Artillery Trends*, July 1963, 4–5; David E. Ott, *Field Artillery: 1954–1973* (Washington, DC: Vietnam Studies, Department of the Army, 1975), 51–53.

42. Ott, *Field Artillery*, 157–59; Lloyd J. Picou, "Airmobile Artillery in Combat," *Artillery Trends*, August 1967, 20; Dastrup, *King of Battle*, 277–78.

43. Dastrup, *King of Battle*, 284, 286.

44. Bailey, *Firepower and Field Artillery*, 12, 53, 60, 69, 253–55.

45. John Gordon IV, "Evolution of Soviet Fire Support, 1940–1988," *Field Artillery*, June 1988, 18–21; Bailey, *Field Artillery and Firepower*, 275–76; Raphael J. Hallada,

"Fire Support Modernization: A Major Step Toward Deterrence," *Military Review*, August 1989, 4–14; Bidwell, *Brassey's Artillery of the World*, 108.

46. *Jane's Armour and Artillery, 1979–1980* (London: Jane's Publications), Foreword; Hogg, *Artillery 2000*, 78–80.

47. Christopher Campbell, *Nuclear Weapons Fact Book* (Novato, CA: Presidio Press, 1984), 130–59.

48. Anthony G. Pokorny, "Take the Tech," *Field Artillery Journal*, September-October 1984, 21–22. The U.S. Army is currently working on submunitions for the Army Tactical Missile System that could neutralize heavily armored vehicles, such as self-propelled artillery and main battlefield tanks, with the objective of having the submunitions fielded late in the 1990s.

49. Bailey, *Field Artillery and Firepower*, 288–89; R.C.F. Craven, "A 70 for the 90s," *Field Artillery Journal*, May-June 1983, 26–31.

50. Bailey, *Field Artillery and Firepower*, 305–6.

51. Hogg, *Artillery 2000*, 108–9.

52. Foss, *Artillery of the World*, 19–24; *Jane's Armour and Artillery, 1979–1980*, Foreword; James F. Hall, "Precision Guided Artillery: First and Second Generation Projectiles," *Field Artillery Journal*, May-June 1981, 9–12; Hogg, *Artillery 2000*, 108–110.

53. Hall, "Precision Guided Artillery," 9.

54. Foss, *Artillery of the World*, 19–24; Hall, "Precision-Guided Artillery," 9–12; Hogg, *Artillery 2000*, 108.

55. Hall, "Precision-Guided Artillery," 9; Bill Rittenhouse, "MLRS Smart Munitions," *Field Artillery*, August 1987, 46.

56. Rittenhouse, "MLRS Smart Munitions," 46–48.

57. Hall, "Precision Guided Artillery," 11–12; Hogg, *Artillery 2000*, 109.

58. Hogg, *Artillery 2000*, 132.

59. Ibid.

60. Bailey, *Field Artillery and Firepower*, 12, 60, 69, 293, 303, 304, 323; Hogg, *Artillery 2000*, 130–32.

61. Matthew J. Ringer and Martell D. Fritz, "FADAC Computations versus Manual Computations," *Artillery Trends*, April 1969, 40; Bidwell, *Brassey's Artillery of the World*, 154; Hogg, *Artillery*, 122.

62. Hogg, *Artillery 2000*, 122–25.

63. Bailey, *Field Artillery and Firepower*, 292; Foss, *Artillery of the World*, 173, 187; Hogg, *Artillery 2000*, 120–24.

64. Hogg, *Artillery 2000*, 291–92, 303–5.

65. Ibid., 144.

66. John K. Cooley, *Payback: America's Long War in the Middle East* (Washington, DC: Brassey's Inc., 1991), 205–25; Barry R. McCaffrey, "Field Artillery Desert Facts," *Field Artillery*, October 1991, 2–3; Vollney B. Corn, Jr., and Richard A. Lacquemont, "Silver Bullets," *Field Artillery*, October 1991, 10–18; Kenneth P. Graves, "Steel Rain: XVIII Airborne Corps Artillery in Desert Storm," *Field Artillery*, October 1991, 49–56; L. Scott Lingamfelter, "In the Wake of a Storm: Improving the FA after Operation Desert Storm," *Field Artillery*, August 1991, 27–29.

Part II

Bibliographical Essays

5

Early Artillery Developments, 1350–1763

Over the years scholars have written about virtually every aspect of the evolution of field artillery. Their works give the reader a solid understanding of changes in materiel, tactics, organization, and operations from the early bombards of the fourteenth century to the nuclear guided missiles and tube artillery of the late twentieth century. Because the Europeans, especially the Germans, French, British, and Russians, and later the Americans have dominated field artillery developments, the bibliographical essays in Part II concentrate on books, articles, theses, and dissertations covering the history of European and American field artillery, with works covering other countries included as appropriate. Part II is divided into four chapters that correlate with the chapters of Part I and covers the major historical and technical publications written in English, with foreign-language publications being considered only if they make a significant contribution.

GENERAL HISTORIES

To understand artillery from 1350 to 1763, the student or scholar requires a firm grasp of the history of warfare. An ambitious work, *War through the Ages* (119) by Lynn Montross, offers a significant examination of the history of warfare. It covers technology, tactics, strategy, materiel, and operations. Archer Jones's *The Art of War in the Western World* (100) is another vital work and a basic source on the history of warfare. Beginning with the ancient wars of the Greeks and continuing his discussion to the Cold War, Jones traces the evolution of tactics, logistics, strategy, technology, and operations and provides vital insights into their interrelationships. Sydney F. Wise and Richard A. Preston's *Men in Arms: A History of Warfare and Its Interrelationships with Western*

Society (182) not only has undergone several different editions since its first publication in the 1950s but also presents a valuable examination of warfare beginning with the wars of the Greeks through the Cold War of the twentieth century, along with an extensive bibliography. J.F.C. Fuller has written several key works on military history. Any serious student of military history should read his *A Military History of the Western World* (64), *Decisive Battles: Their Influence upon History and Civilization* (66), and *The Decisive Battles of the Western World* (67) for their solid interpretations of war and its impact. D. H. Zook and Robin Higham's *A Short History of Warfare* (184) also furnishes a solid analysis on warfare.

Covering a shorter time period than the previous studies cited, William H. McNeill's *The Pursuit of Power: Technology, Armed Forces, and Society since* A.D. *1000* (116) addresses the balance among technology, armed forces, and society, furnishes an invaluable analysis of the development of siege, garrison, and field artillery, and is a landmark work that must be read. Theodore Ropp's *War in the Modern World* (149) is a classic study. It offers invaluable insights into warfare beginning in the fifteenth century and ending with the twentieth century, has excellent content and reference footnotes to guide the student to additional books, and is essential reading. Less scholarly than McNeill's and Ropp's studies but still a solid work, Jac Weller's *Weapons and Tactics: Hastings to Berlin* (178) covers the history of warfare with an emphasis on weapons and tactics, as the title suggests, and furnishes vital information on artillery. Another particularly vital work is Hans Delbrück's multivolume *History of the Art of War* (38), being translated by Walter J. Renfroe. Of the four volumes translated so far, volumes three and four are helpful for their discussions on Gustavus Adolphus (q.v.), Gribeauval (q.v.), and Frederick the Great (q.v.). Volume three also has a sound chapter on the rise of gunpowder artillery. B. H. Liddell Hart's *Strategy: The Indirect Approach* (76) is another critical work.

For general histories that focus upon technology see T. Wintringham's thought-provoking *Weapons and Tactics* (181). Trevor N. Dupuy's *The Evolution of Weapons and Warfare* (51) is a solid account of the evolution of weapons and makes some attempt to quantify the impact of weapons upon tactics, but it might not have an appeal to the serious scholar. J.F.C. Fuller's *Armament and History: A Study of the Influence of Armament on History from the Dawn of Classical Warfare to the Second World War* (65) also gives insights into the impact of technology and artillery on battle. For an interesting work see Robert L. O'Connell's *Of Arms and Men: A History of War, Weapons, and Aggression* (130). It is a critical source for the impact of technology on warfare and has a particularly interesting chapter on the rise of gunpowder artillery. John A. Lynn's (ed.) *Tools of War: Instruments, Ideas, and Initiations of War, 1445–1871* (111) charts how weapons changed warfare, military thought, and the organization of military forces.

Other general histories of warfare are available. Although they are dated, Robert G. Albion's *Introduction to Military History* (2), Arthur Birnie's *The Art of War* (13), and G.F.R. Henderson's *The Science of War* (80) are worth reading for their perceptive interpretations. Oliver Spaulding, Hoffman Nickerson, and John W. Wright's *Warfare: A Study of Military Methods from the Earliest Times* (163) should also be read for its solid appraisal of warfare and attention to detail.

Two key works on the history of military thought are available. Although Edward Mead Earle's (ed.) *Makers of Modern Strategy: Military Thought from Machiavelli to Hitler* (52) is old, it should be considered a starting point for those interested in the history of military thought. Peter Paret's (ed.) follow-on *Makers of Modern Strategy* (135) should also be examined for its perceptive articles on military thought.

Bibliographical reference books are available to guide the student or scholar. Robin Higham's *Official Histories: Essays and Bibliographies from Around the World* (86) and *A Guide to the Sources of British Military History* (84) are excellent sources. Also of value is Gerald Jordan's *British Military History: A Supplement to Robin Higham's Guide to the Sources* (100A). Robin Higham's *A Guide to the Sources of United States Military History* (85) is essential. Robin Higham and Donald Mrozek's *A Guide to the Sources of United States Military History: Supplement I* (87), *A Guide to the Sources of United States Military History: Supplement II* (88) and *A Guide to the Sources of United States Military History: Supplement III (88A)* are additional solid bibliographical works.

MEDIEVAL WARFARE

For an understanding of the conditions that led to the appearance of gunpowder artillery and its impact upon warfare, the student or scholar should also investigate warfare in medieval Europe. John Beeler's *Warfare in Feudal Europe* (9) is a point of departure for any serious study of that period. Beeler's *Warfare in England: 1066–1189* (8) is another critical work to be considered. Terence Wise's *Medieval Warfare* (183) is a sound study. Other works on warfare in medieval Europe include H. W. Koch's illustrated *Medieval Warfare* (102), which has a good section on artillery, Vesey Norman's *Arms and Armour* (122), and Vesey Norman and D. Pottinger's *Warrior to Soldier* (124) and *English Weapons and Warfare: 449–1660* (123).

Any serious student must read Sir Charles Oman's classic study, *The Art of War in the Middle Ages* (132), which has gone through several reprintings. Oman's *A History of the Art of War in the Sixteenth Century* (131) is vital for its discussion of artillery as well as military history in general. Steven Runciman's *History of the Crusades* (151) is a thorough book on medieval warfare. The reader should also consult J. F. Verbruggen's *The Art of Warfare in Western Europe during the Middle Ages* (174). Perhaps the best recent work on warfare in medieval Europe is Philippe Contamine's *War in the Middle Ages* (33). It is a crucial study with a solid bibliography on gunpowder artillery.

Other substantial works cover the early years of gunpowder artillery also. The reader should consult Charles H. Ashdown's *Armor and Weapons in the Middle*

Ages (3). See Fredric J. Baumgartner's *From Spear to Flintlock: A History of War in Europe and the Middle East to the French Revolution* (4) for a short discussion on early artillery. Although it covers more than just the Middle Ages, A. Corvisier's *Armies and Societies in Europe: 1494–1789* (35) is a landmark study for its perceptive analysis of the relation between armies and society. Geoffrey Parker's *The Military Revolution: Military Innovation and the Rise of the West, 1500–1800* (136) is particularly valuable when considering the role of artillery during the latter years of the Middle Ages. Also see Fletcher Pratt's "Marignano: The First Artillery Battle" (140) for a critical analysis of a key battle.

WARFARE IN THE SEVENTEENTH AND EIGHTEENTH CENTURIES

There are several critical books on warfare in the seventeenth and eighteenth centuries for the reader to examine. David Chandler's *The Art of Warfare in the Age of Marlborough* (25) and Christopher Duffy's *The Military Experience in the Age of Reason* (49) are reliable books for an introduction to the topic. Chandler's study is a classic that must be consulted, while Duffy's book furnishes vital insights into the impact that the Age of Reason had on warfare and artillery. Although Brent Nosworth's *The Anatomy of Victory: Battle Tactics, 1689–1763* (128) is more appealing to the general reader than the scholar, he scrutinizes tactics, organization of armies, and equipment, and is worth reading. For another penetrating book on warfare from the seventeenth century to the early nineteenth century see Russell F. Weigley's *The Age of Battles: The Quest for Decisive Warfare from Breitenfeld to Waterloo* (177). He discusses the quest for the decisive battle and finds that it rarely occurred. Robert S. Quimby's *The Background of Napoleonic Warfare: The Theory of Military Tactics in Eighteenth Century France* (141) is a key work for understanding the foundations of mid-eighteenth-century warfare and is a must for his solid analysis of artillery materiel, tactics, and organization. Although it is a brief work, Anthony Kemp's *Weapons and Equipment of the Marlborough Wars* (101) is worth examining. Its chapter on artillery materiel, tactics, and organization is important. For a critical battle see Fletcher Pratt's "Breitenfeld: The Battle That Changed the World" (139).

ARTILLERY HISTORIES

General histories on artillery are available for studying. Joseph Jobe's (ed.) *Guns: An Illustrated History of Artillery* (99) is a well-written, illustrated general history of artillery. However, it lacks footnotes. Although it is less scholarly than many accounts of the development of artillery, Phillip H. Stevens's *Artillery through the Ages* (164) gives the reader a sound overview of the history of artillery. O.F.G. Hogg in *Artillery: Its Origin, Heyday, and Decline* (92) explores the evolution of artillery materiel, especially English, and is a standard work on the topic. Written for the U.S. National Park Service, Albert Manucy's short *Artillery through the Ages: A Short Illustrated History of Cannon, Emphasizing Types Used in America* (113) covers artillery, gunpowder, characteristics of cannons, and projectiles from the sixteenth through the nineteenth century in America. John Batchelor and Ian V. Hogg's *Artillery* (7) is a solid illustrated history of the development of artillery but lacks footnotes and a sound bibliography. Although Frank E. Comparato's *Age of Great Guns: Cannon Kings and Cannoneers Who Forged the Firepower of Artillery* (32) is confusing in parts because of faulty organization, it has a wealth of technical information on artillery from its beginnings through the mid–1960s and is perhaps the best general work available on the history of artillery.

Even though it is less scholarly than any of the other books, Fairfax Downey's *Cannonade: Great Artillery Actions of History, the Famous Cannons, and the Master Gunners* (44) is interesting reading because of its moving and lively accounts of battles and individuals, and it has a solid bibliography. Although William Reid's *Arms through the Ages* (143) is a comprehensive work, it is useful on early arms and artillery as is Auguste Demmins *An Illustrated History of Arms and Armour* (39). A short primer on artillery is the U.S. Army Field Artillery School's *History of the Development of Field Artillery Materiel* (170). Written for U.S. Army field artillerymen, it covers all the basic developments in field artillery history from the initial bombards through 1940.

More specialized works on the beginnings of artillery are also available for consulting. Henry W. L. Hime's *The Origin of Artillery* (89) is one of the first scholarly accounts and is a standard work on the subject. J. R. Portington's *A History of Greek Fire and Gunpowder* (136) also covers early artillery. M.G.A. Vale's "New Techniques and Old Ideals: The Impact of Artillery on War and Chivalry at the End of the Hundred Years War," in C. T. Allmand's (ed.) *War, Literature, and Politics in the Late Middle Ages* (171) is imperative for understanding the transition from trebuchets and ballistae to gunpowder artillery and has valuable footnotes of English and non-English sources for the period. Sir Henry Brackenbury's "Ancient Cannon in Europe" (17) should be read for its useful examination of the first gunpowder artillery in Europe. Robert C. Clephan's *The Ordnance of the Fourteenth and Fifteenth Centuries* (29) has a solid discussion on bombards and other early gunpowder artillery. Another work by Clephan is "Early Ordnance of Europe" (28). H. Schubert's "The First Cast-Iron Cannon Made in England" (157) is also an important study, as is his *History of British Iron and Steel Industry from c. 450* B.C. *to* A.D. *1775* (158) for its discussion of early artillery. R. J. Forbes's *Man the Maker: A History of Technology and Engineering* (61) examines metallurgy among other technical subjects and discusses the rise of cast iron and cast-iron artillery in Europe in a short chapter. A short work with an outstanding bibliography, John M. Patrick's *Artillery and Warfare during the Thirteenth and Fourteenth Centuries* (137) provides a thoughtful analysis of the beginnings of gunpowder artillery and its influence on warfare in Europe. C. S. Forester's *The Gun* (62) has an excellent section on early artillery.

Carlo M. Cipolla in *Guns, Sails, and Empires: Technological Innovation and the Early Phases of European Expansion, 1400–1700* (27) has one of the most incisive accounts of the early development of artillery in Europe and should be consulted whenever researching early artillery. J. R. Hale's "Gunpowder and the Renaissance" (74) covers the impact of gunpowder artillery on siege warfare in the fourteenth and fifteenth centuries. The author also covers the opposition to gunpowder artillery. A. Essenwein's *Quellen zur Geschichte der Feuerwaffen* (Sources on the History of Firearms, 54) also examines early European artillery as well as small arms, while Theodore Wertime in *The Coming of the Age of Steel* (179) has a short but worthwhile section on artillery. For other valuable discussions on early artillery see A. W. Wilson's *The Story of the Gun* (180) and Kelly DeVries's *Medieval Military Technology* (41).

Interesting works on artillery during the late Middle Ages also exist. They include L. N. Bonaparte (Napoleon III) and M. Favé's six-volume study, *Études sur le Passé et l'Avenir de l'Artillerie* (Studies on the Past and Future of Artillery, 15). The first volume covers the years from 1328 to 1815. Other books are J. Hayward's *The Art of the Gunmaker* (79) and V. Schmidtchen's *Bombarden, Befestigungen, Büchsenmeister, von den ersten Mauerbrechern des Spätmittelalters zur Belagerungsartillerie der Renaissance: eine Studie zur Entwicklung der Militärtechnik* (Bombards, Fortifications, and Master Armors from the First Siege Artillery of the Late Middle Ages to the Siege Artillery of the Renaissance: A Study of the Development of Military Technology, 155). Some other works worth reading are W. W. Greener's *The Gun and Its Development* (73) and Harold Lamb's *The Earth Shakers* (107). An old study on the topic is M. P. LaCombe's *Arms and Armour in Antiquity and the Middle Ages* (106), translated by Charles Boutell. LaCombe's book has a lengthy chapter that covers the rise of artillery through the mid-nineteenth century. Bernard Rathgen's *Das Geschütz im Mittelalter* (Ordnance of the Middle Ages, 142) is also worthwhile. An older but still useful work is G. Hue's *L'Artillerie dans l'Antiquité et au Moyen Age* (Artillery from Antiquity to the Middle Ages, 95). Rudolf Schneider's "Die Artillerie des Mittelalters" (Artillery of the Middle Ages, 156) is critical. Boyd L. Dastrup's *King of Battle: A Branch History of the U.S. Army's Field Artillery* (37) has a short section on early artillery. Henry W. L. Himes's "Who Invented the Leather Guns?" (90) explores the origins of leather artillery pieces during the early 1600s. For an insightful look at Spanish artillery in the fifteenth century see Westen F. Cook, Jr's. "The Cannon Conquest of Nāsrid Spain and the End of the Reconquista" (34A).

TECHNOLOGY

For those interested in technology and artillery, Martin Van Creveld's *Technology and War: From 2000 B.C. to the Present* (173) is one of the latest works.

Even though some of his conclusions are questionable and not substantiated with footnotes, a student of military history should read his book for his perceptive analysis and bibliography. Bernard and Fawn M. Brodie's *From Crossbow to H-Bomb* (19) is the standard work on the history of technology and war that should be in every student's or scholar's library. Melvin Kranzberg and Carroll W. Pursell, Jr.'s (eds.) seven-volume work, *Technology in Western Civilization* (104), and Charles Singer et al.'s (eds.) *A History of Technology* (161) are solid reference works.

More specialized studies on the history of technology are available. A. R. Hall in *Ballistics in the Seventeenth Century* (75) covers the state of ballistics in the seventeenth century in Europe and has written one of the best recent works on the topic. Besides exploring gunnery and ballistics, Hall examines the contributions of seventeenth-century artillery experts, such as Niccolò Tartaglia and Luis Collado. Benjamin Robins's *The New Principles of Gunnery* (148) is one of the first books on scientific ballistics and is, therefore, a critical study. For an early study of the history of gunnery see James Glenie's *A History of Gunnery* (71).

Various historians cover technology of a specific country. Charles Ffoulkes's *The Gun-Founders of England* (58) is an essential work. He covers the basic development of cannon manufacturing from its beginning through the first decades of the eighteenth century and includes a list of gun manufacturers in England during those years. Ffoulkes's *Arms and Armament* (57) concentrates on the development of English weaponry from 1300 to 1800 and includes a short but solid chapter on artillery, while his *Armour and Weapons* (56) is also useful for English artillery. For a broader perspective on English technology consult Henry J. Webb's *Elizabethan Military Science: The Books and Practice* (176), which also has an excellent section on artillery tactics, organization, technology, and gunnery. A valuable work on English cannon production is O.F.G. Hogg's *The Royal Arsenal: Its Background, Origin, and Subsequent History* (94). In two volumes Hogg provides an in-depth discussion. Clive Trebilcock's "Spin-off in British Economic History: Armaments and Industry, 1760–1914" (168) writes about British efforts to construct a cannon-boring machine in the 1770s. S. James Gooding's *An Introduction to British Artillery in North America* (72) examines the first artillery brought by the British to North America.

Violet Barbour's *Capitalism in Amsterdam in the 17th Century* (5) and Claude Gaier's *Four Centuries of Liège Gunmaking* (68) cover the development of artillery manufacturing in the Low Countries. P.M.J. Conturie's *Histoire de la Fonderie Nationale de Ruelle, 1750–1940, et des Anciennes Fonderies de Canons de Fer de la Marine* (History of the National Foundry at Ruelle, 1750–1940, and the Ancient Foundries for Cannon for the Navy, 34) discusses cannon manufacturing in France over a period of two centuries. Specifically, the book addresses Jean Maritz's role in building cannon-boring machines during the middle decades of the eighteenth century. Harold L. Peterson's *Round Shot and Rammers* (138) is a standard work on American artillery materiel from the early sixteenth century through the mid-nineteenth century.

From the first days of gunpowder artillery, Europeans have written about the principles of gunnery, the properties of gunpowder, and the art of gun making. Although many of their works are not readily available, students of military and field artillery history should be aware of their existence and contribution to the development of artillery. Roberto Valturio's *De l'Arte Militare* (The Military Art, 172) covers artillery as well as the other combat arms. The founder of the theory of gunnery, Niccolò Tartaglia, in *Nova Scientia* (New Knowledge, 165) deals with the composition of gunpowder, gives illustrations of guns and mortars, and outlines the duties of a gunner. Because of the importance of Tartaglia's work, many European authors of the sixteenth century based their work on it. William Bourne's *The Arte of Shooting Great Ordnance, containing very necessary matters of all sorts of seruitours eyther by Sea or Land* (16) reflects the influence of Tartaglia on the author. *The Handbook of Artillery* (31), by Luis Collado, an engineer in the Spanish army in Italy during the reigns of Philip II and Philip III of Spain, is the first detailed and well-illustrated technical manual on artillery in Spanish and was later translated into English. In his book Collado discusses the theory and practice of artillery and examines the various uses of artillery. Don Diego Uffano's *Treatise on Artillery and Its Use* (169) explains founding, management, supply, and tactical use of artillery. Henry Hexham's *The Third Part of the Principles of the Art Military* (82) is another book on early cannon.

Books on early gunnery are worth examining. Although he borrows extensively from Tartaglia and Uffano, Robert Norton's *The Gunner shewing the Whole Practice of Artillery* (127) is an early work on gunnery and other artillery topics. Norton's *Of the Art of Great Artillery* (126) is another solid study on gunnery. *Gunners Dialogue with the Art of Great Gunnery* (125) is another book by Robert Norton. John Roberts's *The Compleat Cannoniere* (146) is especially good for dimensions, charges, and ranges. Thomas Smith's *The Art of Gunnery* (162) is also worth reading for those interested in early gunnery. Those interested in gunnery in that same century should consult Master Gunner Nye's *Art of Gunnery* (129). Besides these, John Muller's *A Treatise on Artillery* (121) is a basic source on English artillery technology in the eighteenth century and is a must for understanding British artillery reforms of the last decades of the century. Reprinted several times in English, French, Italian, and German, Casimir Simienowicz's *The Great Art of Artillery* (159) is a sound source for European artillery in the seventeenth century.

Other early works include William Eldred's *The Gunner's Glasse* (53) and Tomas Moretti's *A General Treatise of Artillery* (120), translated by Sir Jonas Moore. Thomas Binning's *A Light to the Art of Gunnery* (12), reprinted several times, with the last reprinting in 1744, discusses cannon design, manufacture, and gunnery. Michael Meithem's *German Treatise on Artillery* (117) is also worth reading.

Books on early gunpowder artillery in other European languages appeared at the same time as these English works did. Gabriel Busca's *Instruttione de Bom-*

bardieri (Instruction for Artillerymen, 22) contains some information on gunnery. Another early Italian work on artillery is Eugenio Gentilini's *Instruttione de Bombardiere* (Instruction for Artillerymen, 69). Pietro Sardi's three-volume *L'Artiglierìa* (The Artillery, 153) deals with artillery as well as gunpowder, while Menno van Coehorn's *Instruction sur le Fait de L'Artillerie* (Instruction on the Facts of Artillery, 30) is useful for its discussion of cannon construction. M. Bellidor's *Le Bombardier Français* (The French Bombardier, 11) is a solid source for French artillery technology. Français Blondel's *L'Art de Jeter les Bombes* (The Art of Throwing Bombs, 14) and D. Rivault's *Les Élémens de l'Artillerie* (Elements of Artillery, 145) are solid sources. M. Surirey de Saint-Rémy's *Mémoire d'Artillerie* (Treatise on Artillery, 152) is a standard source on French artillery materiel in the seventeenth and early eighteenth centuries. This work appeared in three principal editions between 1697 and 1747. Each edition was revised to reflect the latest developments in French artillery. T. Cerda's *Lecciónes de Artilleria* (Lessons in Artillery, 24) is also vital for early artillery technology in Europe. The latest edition (1747) includes information on the Vallière system. Wilhelm Hassenstein's (ed.) *Das Feuerwerksbuch von 1420* (The Book on Fireworks from 1420, 77) contains pictures on cannon from the early sixteenth century. F. A. Geszeler's *Das schweizerische Geschühtzwesen zur Zeit des Schwaben krieges* (Swiss Ordnance during the Swabian War, 70) is sound. An essential book on American and European artillery is Louis de Tousard's two-volume *American Artillerist's Companion* (167). It is particularly valuable for the information on European artillery of the mid-eighteenth century.

TACTICS, DOCTRINE, AND ORGANIZATION

Because artillery served its earliest duty as a siege weapon, students of military history should at least examine siege warfare. Both Christopher Duffy's *Siege Warfare: The Fortress in the Early Modern World, 1494–1660* (45) and its accompanying volume, *The Fortress in the Age of Vauban and Frederick the Great, 1660–1789* (48), are solid, detailed studies with extensive bibliographies. *Siege Warfare* covers early cannon and siegecraft and the conversion of castles and fortresses to withstand siege cannon. Less scholarly but still valuable is Ian V. Hogg's *Fortress: A History of Military Defense* (91). It is an illustrated book on castles and siege warfare. Recently published, Martin Brice's *Stronghold: A History of Military Architecture* (18) charts the history of fortifications from 3500 B.C. to 1945. Bryan O'Neill's *Castles and Cannon: A Study of Early Artillery Fortifications in England* (133) should be examined when studying fourteenth century castles and artillery. For a contemporary look at siegecraft and fortresses see Gabriel Busca's *Della Espugnatione et Difesa delle Fortezze* (Storming and Defending the Fortress, 21) for its numerous clear and illuminating diagrams showing engineering methods of the sixteenth century.

For artillery tactics and operations on the battlefield see B. P. Hughes's *Firepower: Weapon Effectiveness on the Battlefield, 1630–1850* (97). He examines field artillery drill and tactics, the theoretical performance of artillery, and musket drill, and has produced one of the finest works on firepower. He has also written *British Smoothbore Artillery: The Muzzle Loading Artillery of the 18th and 19th Centuries* (96) and *Open Fire: Artillery Tactics from Marlborough to Wellington* (98). While both studies cover much of the same material, they are essential reading for smoothbore artillery tactics and organization. Hughes's books should form a core of reading on tactics and organization during the height of smoothbore artillery. Although Gunther E. Rothenberg's *The Art of Warfare in the Age of Napoleon* (150) discusses warfare in general in the eighteenth and nineteenth centuries, he also has excellent sections on field artillery materiel, tactics, and organization prior to Napoleon. Sir Edward S. May's *Achievements of Field Artillery* (115), in contrast, focuses on the impact of artillery on battlefield operations from the mid-seventeenth to the mid-nineteenth century and provides exacting detail of field artillery operations in combat. Philip J. Haythornthwaite's well-illustrated *Weapons and Equipment of the Napoleonic Wars* (78) is a must to read for its discussion on artillery from the mid-eighteenth century through the early nineteenth century.

For a more narrow focus that examines artillery of a specific country see Kenneth Brookes's *Battle Thunder: The Story of Britain's Artillery* (20) for a rich, detailed standard work on British artillery history from the fourteenth century through the late twentieth century. In *English Artillery: 1326–1716* (93) O.F.C. Hogg supplies a solid account of English artillery before the organization of the Royal Artillery Regiment in in 1716. F. L. Taylor in *The Art of War in Italy, 1494–1529* (166) provides valuable insights into siege artillery and the French campaign of 1494, in which the French introduced some of the first relatively mobile cannon for employment in the field. Christopher Bellamy's *Red God of War: Soviet Artillery and Rocket Forces* (10) is an excellent history of Russian and Soviet artillery and the contributions of the Russians during the eighteenth century. Moreover, Bellamy's bibliography contains an outstanding list of Russian and English sources. Christopher Duffy's *The Army of Maria Theresa: The Armed Forces of Imperial Austria, 1740–1780* (46) covers the Austrian army as well as exploring the Liechtenstein (q.v.) artillery reforms of the early eighteenth century in organization, ordnance, and equipment. Another solid work on Austrian field artillery tactics, materiel, and organization is A. Dolleczek's *Geschichte der österreichischen Artillerie* (History of Austrian Artillery, 43). Although it is old and hard to obtain, the book is informative and useful for its analysis of Austrian artillery. Chevalier d'Arcy's *Observations et Expériences sur l'Artillerie* (Observations and Experiences of the Artillery, 36) covers French artillery operations in the eighteenth century. For Belgian artillery consult Paul Henrad's *Histoire de l'Artillerie en Belgique* (The History of Belgian Artillery, 81).

For valuable insights into Prussian field artillery tactics during the eighteenth century see Frederick the Great's *Die politischen Testaments Friedrichs des Grossen* (The Political Testament of Frederick the Great, 63). Richard Fester's (ed.) *Die Instruktion Friedrichs des Grossen für seine Generale* (Frederick the Great's Instructions for his Generals, 55) is worth examining. Edited and translated by Jay Luvaas, *Frederick the Great on the Art of War* (110) offers vital insights into field artillery tactics during the mid-eighteenth century. Of equal importance are the general works on warfare of the mid-eighteenth century that cover Prussian artillery tactics. Although J. Mauvillon's *Geschichte Ferdinands Herzogs von Braunschweig-Lüneburg* (The History of Ferdinand, the Duke of Braunschweig-Lüneburg, 114) encompasses military campaigns and operations as a whole, he also furnishes invaluable insights into artillery operations during the Seven Years' War (1756–1763). H. G. Mirabeau and J. Mauvillon's *Système Militaire de Prussse* (The Military System of Prussia, 118) is another old work but a critical one on the development of Prussian artillery in the mid-eighteenth century. For Prussian artillery of the mid-eighteenth century, consult *Über die Wirkung des Feuergewehrs* (The Operation of Firearms, 154), by Gerhart von Scharnhorst, a Prussian general noted for his reform efforts early in the nineteenth century. C. F. Barsewisch's *Meine Kriegs-Erlebnisse während des siebenjährigen Krieges, 1757–1763* (My War Experiences during the Seven Years' War, 1757–1763, 6) supplies solid insights into artillery tactics, especially Prussian, during the Seven Years' War.

Books on leading artillery personalities are also good sources for artillery tactics, organization, and materiel. F. G. Heyman's *John Ziska and the Hussite Revolution* (83) is solid. Works on Gustavus Adolphus include Michael Roberts's *Gustavus Adolphus: A History of Sweden, 1611–1632* (147), which is the standard work on the king as is Ahlund Nils's *Gustav Adolf the Great* (1). Sir George MacMunn's *Gustavus Adolphus: The Northern Hurricane* (112) is another book on the famous king. A more recent work, C.R.L. Fletcher's *Gustavus Adolphus* (59), should also be examined for its analysis of the king's operations and artillery, while Theodore A. Dodge's *Gustavus Adolphus* (42) is another solid work. B. Chapman's *The History of Gustavus Adolphus* (26) is an earlier but sound work. J. Watts De Peyster's *The History of the Life of Leonard Tortennson* (40) covers the life of one of Sweden's most important artillerymen in the seventeenth century. For a study on the French artilleryman, who was critical in developing French artillery during the sixteenth century consult Francois de Vaux Foletier's *Galiot de Genouillac: Maître de l'Artillerie de France* (Galiot de Genouillac: The Artillery Master of France, 60).

Several valuable works are available on Frederick the Great that should be considered when examining Prussian field artillery during the mid-eighteenth century. Gerhard Ritter's *Frederick the Great: A Historical Profile* (144), translated by Peter Paret in 1968, is a classic study on the king. Although it is an old book, Thomas Carlyle's mammoth study, *History of Frederick the Great* (23), should be consulted. More recent is Franz Kugeler's *Geschichte Friedrichs des Grossen* (The History of Frederick the Great, 105). Christopher Duffy's *The Military Life of Frederick the Great* (50) and *The Army of Frederick the Great* (47) should be read for their insights not only into Frederick's army but also his artillery. Edith Simon's *The Making of Frederick the Great* (160) is also worth examining.

Vauban has been the subject of several books. One of the best is P. E. Lazard's *Vauban, 1666–1703* (108).

For drill manuals consult J. B. Koch's *Artilleristen Hand-Buch* (The Gunner's Handbook, 103). It is an example of early manuals for artillerymen, as are H. Vogel's *Kurzer Bericht der Artillerie-Wissenschaft* (A Short Report on the Science of Artillery, 175) and M. LeBlond's *Élémens de Tactique* (The Elements of Tactics, 109). M. Bellidor's *Le Bombardier Français* (The French Bombardier, 11) examines French artillery drill. These manuals on field artillery along with historical works are valuable for their insights on tactics of the mid-eighteenth century.

BIBLIOGRAPHY

1. Ahlund, Nils. *Gustav Adolf the Great*. Translated by Michael Roberts. Princeton, NJ: Princeton University Press, 1940.
2. Albion, Robert G. *Introduction to Military History*. New York: Century, 1929.
3. Ashdown, Charles H. *Armor and Weapons in the Middle Ages*. United Kingdom: New Holland Publications, 1925.
4. Baumgartner, Fredric J. *From Spear to Flintlock: A History of War in Europe and the Middle East to the French Revolution*. New York: Praeger, 1991.
5. Barbour, Violet. *Capitalism in Amsterdam in the 17th Century*. Reprint. Ann Arbor, MI, University of Michigan Press, 1963.
6. Barsewisch. C. F. *Meine Kriegs-Erlebnisse während des siebenjährigen Krieges, 1757–1763*. Berlin, 1863.
7. Batchelor, John, and Ian V. Hogg. *Artillery*. New York: Charles Scribner's Sons, 1972.
8. Beeler, John. *Warfare in England: 1066–1189*. Ithaca, NY: Cornell University Press, 1966.
9. ———. *Warfare in Feudal Europe*. Ithaca, NY: Cornell University Press, 1972.
10. Bellamy, Christopher. *Red God of War: Soviet Artillery and Rocket Forces*. Washington, DC: Brassey's Defence Publishers, 1986.
11. Bellidor, M. *Le Bombardier Français*. Paris, 1731.
12. Binning, Thomas. *A Light to the Art of Gunnery*. London, 1689.
13. Birnie, Arthur. *The Art of War*. London: Nelson, 1942.
14. Blondel, François. *L'Art de Jeter les Bombes*. Paris, 1683.
15. Bonaparte, Louis Napoleon (Napoleon III), and M. Favé. *Études sur le Passé et l'Avenir de l'Artillerie*. 6 vols. Paris: J. Dumaine, 1846–1871.
16. Bourne, William. *The Arte of Shooting Great Ordnance, containing very necessary matters of all sorts of seruitours eyther by Sea or Land*. London, 1587.
17. Brackenbury, Sir Henry. "Ancient Cannon in Europe." *Proceedings of the Royal Artillery Institute*, 1865–1866.
18. Brice, Martin H. *Strongholds: A History of Military Architecture*. New York: Schocken Books, 1985.

19. Brodie, Bernard and Fawn. *From Crossbow to H-Bomb*. Bloomington, IN: Indiana University Press, 1973.
20. Brookes, Kenneth. *Battle Thunder: The Story of Britain's Artillery*. Berkshire, England: Osprey Publishing Ltd., 1973.
21. Busca, Gabriel. *Della Espugnatione et Difesa delle Fortezze*. Turin, 1585.
22. ———. *Instruttione de Bombardiere*. Turin, 1598.
23. Carlyle, Thomas. *History of Frederick the Great*. 10 vols. London: Chapman and Hall, 1858–1865.
24. Cerda, T. *Lecciónes de Artilleria*. Madrid, 1644.
25. Chandler, David. *The Art of Warfare in the Age of Marlborough*. New York: Hippocrene Books, 1976.
26. Chapman, B. *The History of Gustavus Adolphus*. London: Longman, Brown, Green, and Longmans, 1856.
27. Cipolla, Carlo M. *Guns, Sails, and Empires: Technological Innovation and the Early Phases of European Expansion, 1400–1700*. Reprint. Manhattan, KS: Sunflower University Press, 1985.
28. Clephan, Robert C. "Early Ordnance of Europe." *Archaeologia Aliana* XXV (1903).
29. ———. *The Ordnance of the Fourteenth and Fifteenth Centuries*. No publisher, ca. 1900.
30. Coehorn, Menno van. *Instruction sur le Fait de L'Artillerie*. Paris, 1633.
31. Collado, Luis. *The Handbook of Artillery*, Venice, 1586.
32. Comparato, Frank E. *Age of Great Guns: Cannon Kings and Cannoneers Who Forged the Firepower of Artillery*. Harrisburg, PA: The Stackpole Company, 1965.
33. Contamine, Philippe. *War in the Middle Ages*. Translated by Michael Jones. London: B. Blackwell, 1984.
34. Conturie, P.M.J. *Historie de la Fonderie Nationale de Ruelle, 1750–1940, et des Anciennes Fonderies de Canon de Fer de la Marine*. Paris, 1951.
34A. Cook, Weston F., Jr. "The Cannon Conquest of Nāsrid Spain and the End of the Reconquista." *Journal of Military History*, January 1993, 43–70.
35. Corvisier, A. *Armies and Societies in Europe: 1494–1789*. Translated by Abigail T. Siddall. Bloomington, IN: Indiana University Press, 1979.
36. D'Arcy, Chevalier. *Observations et Expériences sur l'Artillerie*. 1755.
37. Dastrup, Boyd L. *King of Battle: A Branch History of the U.S. Army's Field Artillery*. Fort Monroe, VA: Office of the Command Historian, U.S. Army Training and Doctrine Command, 1992.
38. Delbrück, Hans. *History of the Art of War*. Translated by Walter J. Renfroe. Westport, CT: Greenwood Press, 1975.
39. Demmin, Auguste. *An Illustrated History of Arms and Armour*. London, 1894.
40. De Peyster, J. Watts. *The History of the Life of Leonard Tortennson*. Poughkeepsie, NY: Platt and Schram, 1955.
41. DeVries, Kelly. *Medieval Military Technology*. Peterborough, Ontario: Broadview Press, 1992.
42. Dodge, Theodore. *Gustavus Adolphus*. Boston: Houghton Mifflin, 1896.
43. Dolleczek, A. *Geschichte der österreichen Artillerie*. Vienna: Author, 1887.
44. Downey, Fairfax. *Cannonade: Great Artillery Actions of History, the Famous Cannons, and the Master Gunners*. Garden City, NY: Doubleday, 1966.

45. Duffy, Christopher. *Siege Warfare: The Fortress in the Early Modern World, 1494–1660*. London: Routledge and Kegan Paul, 1979.
46. ———. *The Army of Maria Theresa: The Armed Forces of Imperial Austria, 1740–1780*. New York: Hippocrene Books, 1977.
47. ———. *The Army of Frederick the Great*. London: David and Charles, 1974.
48. ———. *The Fortress in the Age of Vauban and Frederick the Great, 1660–1789*. London: Routledge and Kegan Paul, 1985.
49. ———. *The Military Experience in the Age of Reason*. New York: Atheneum, 1988.
50. ———. *The Military Life of Frederick the Great*. New York: Atheneum, 1986.
51. Dupuy, Trevor N. *The Evolution of Weapons and Warfare*. Indianapolis: Bobbs-Merrill, 1980.
52. Earle, Edward Mead, ed. *Makers of Modern Strategy: Military Thought from Machiavelli to Hitler*. Princeton, NJ: Princeton University Press, 1943.
53. Eldred, William. *The Gunner's Glasse*. London, 1646.
54. Essenwein, A. *Quellen zur Geschichte der Feuerwaffen*. Leipzig: F. A. Brockhaus, 1877.
55. Fester, Richard, ed. *Die Instruktion Friedrichs des Grossen für seine Generale von 1747*. Berlin, 1946.
56. Ffoulkes, Charles J. *Armour and Weapons*. Oxford, England: Clarendon Press, 1945.
57. ———. *Arms and Armament*. London: Harrap, 1945.
58. ———. *The Gun-Founders of England*. Cambridge, England: Cambridge University Press, 1937.
59. Fletcher, C.R.L. *Gustavus Adolphus*. New York: G. P. Putnam's Sons, 1890.
60. Foletier, François de Vaux. *Galiot de Genouillac, Maître de l'Artillerie de France (1465–1546)*. Paris: Auguste Picard, 1925.
61. Forbes, R. J. *Man the Maker: A History of Technology and Engineering*. London: Constable, 1958.
62. Forster, C. S. *The Gun*. Boston: Little, Brown, 1933.
63. Frederick the Great. *Die politischen Testaments Friedrichs des Grossen*. Berlin, 1920.
64. Fuller, J.F.C. *A Military History of the Western World*. 3 vols. New York: Funk and Wagnalls, 1953–1957.
65. ———. *Armament and History: A Study of the Influence of Armament on History from the Dawn of Classical Warfare to the Second World War*. New York: Charles Scribner's Sons, 1945.
66. ———. *Decisive Battles: Their Influence upon History and Civilization*. New York: Charles Scribner's Sons, 1940.
67. ———. *The Decisive Battles of the Western World*. 3 vols. London: Eyre and Spottiswoode, 1954–1956.
68. Gaier, Claude. *Four Centuries of Liège Gunmaking*. London: Sotheby Park Bernet, 1977.
69. Gentilini, Eugenio. *Instruttione de' Bombardieri*. Venice, 1592.
70. Geszeler, F. A. *Das schweizerische Geschützwesen zur Zeit des Schwabenkrieges*. Zurich: Beer and Company, 1927–1929.
71. Glenie, James. *A History of Gunnery*. Edinburgh, 1786.

72. Gooding, S. James. *An Introduction to British Artillery in North America*. Ottawa, Ontario: Museum Restoration Service, 1965.
73. Greener, W. W. *The Gun and Its Development*. New York: Charles Scribner's Sons, 1897.
74. Hale, J. R. "Gunpowder and the Renaissance." In C. H. Carter, *From Renaissance to Counter-Reformation*. New York, Random House 1966.
75. Hall, A. R. *Ballistics in the Seventeenth Century*. Cambridge, England: Cambridge University Press, 1952.
76. Hart, B. H. Liddell. *Strategy: The Indirect Approach*. New York: Praeger, 1954.
77. Hassenstein, Wilhelm, ed. *Das Feurwerksbuch von 1420*. Reprint. Munich, 1941.
78. Haythornthwaite, Philip J. *Weapons and Equipment of the Napoleonic Wars*. Poole, England: Blandford Press, 1969.
79. Hayward, John F. *The Art of the Gunmaker*. 2 vols. New York: St. Martins, 1962.
80. Henderson, G.F.R. *The Science of War*. London: Longmans Green, 1912.
81. Henrad, Paul. *Histoire de l'Artillerie en Belgique*. Brussels, 1865.
82. Hexham, Henry. *The Third Part of the Principles of the Art Military*. Rotterdam, 1643.
83. Heyman, Frederick G. *John Ziska and the Hussite Revolution*. New York: Russell, 1969.
84. Higham, Robin, ed. *A Guide to the Sources of British Military History*. Berkeley, CA: University of California Press, 1971.
85. ———. *A Guide to the Sources of United States Military History*. Hamden, CT: Archon Books, 1975.
86. ———. *Official Histories: Essays and Bibliographies from Around the World*. Manhattan, KS: Kansas State University Library, 1970.
87. Higham, Robin, and Donald J. Mrozek, eds. *A Guide to the Sources of United States Military History: Supplement I*. Hamden, CT: Archon Books, 1981.
88. ———. *A Guide to the Sources of United States Military History: Supplement II*. Hamden, CT: Archon Books, 1986.
88A. ———. *A Guide to the Sources of United States Military History: Supplement III*. Hamden, CT: Archon Books, 1993.
89. Himes, Henry W.L. *The Origin of Artillery*. London: Green and Company, 1915.
90. ———. "Who Invented the Leather Guns?" *Proceedings of the Royal Artillery*, 1898.
91. Hogg, Ian V. *Fortress: A History of Military Defense*. New York: St Martin's Press, 1975.
92. Hogg, O.F.G. *Artillery: Its Origin, Heyday, and Decline*. Hamden, CT: Archon Books, 1970.
93. ———. *English Artillery: 1326–1716*. London: Royal Artillery Institution, 1963.
94. ———. *The Royal Arsenal: Its Background, Origin, and Subsequent History*. 2 vols. London: Oxford University Press, 1963.
95. Hue, G. *L'Artillerie dans l'Antiquité et au Moyen Age*. Paris, 1880.
96. Hughes, B. P. *British Smoothbore Artillery: The Muzzle Loading Artillery of the 18th and 19th Centuries*. Harrisburg, PA: Stackpole Books, 1969.
97. ———. *Firepower: Weapon Effectiveness on the Battlefield, 1630–1850*. New York: Charles Scribner's Sons, 1974.
98. ———. *Open Fire: Artillery Tactics from Marlborough to Wellington*. Chichester, England: Antony Bird Publications Ltd., 1983.

99. Jobe, Joseph, ed. *Guns: An Illustrated History of Artillery*. Greenwich, CT: New York Graphic Society, 1971.
100. Jones, Archer. *The Art of War in the Western World*. Urbana, IL: University of Illinois Press, 1987.
100A. Jordan, Gerald, ed. *British Military History: A Supplement to Robin Highaoms Guide to the Sources*. London: Garland Publishing, Inc., 1988.
101. Kemp, Anthony. *Weapons and Equipment of the Marlborough Wars*. Poole, England: Blandford Press, 1980.
102. Koch, H. W. *Medieval Warfare*. New York: Cresent Books, 1983.
103. Koch, J. B. *Artilleristen Hand-Buch*. Frankfurt and Leipzig, 1765.
104. Kranzberg, Melvin, and C. W. Pursell, Jr., eds. *Technology in Western Civilization*. 7 vols. New York: Oxford University Press, 1967.
105. Kugeler, Franz. *Geschichte Friedrichs des Grossen*. Leipzig: C. A. Seeman, 1915.
106. LaCombe, M. P. *Arms and Armour in Antiquity and the Middle Ages*. Translated by Charles Boutell. London: Cassell, Petter, and Galpin, 1869.
107. Lamb, Harold. *The Earth Shakers*. Garden City, NY: Doubleday, 1953.
108. Lazard, Pierre Elizier. *Vauban 1666–1703*. Paris, 1934.
109. LeBlond, M. *Élémens de Tactique*. Paris, 1761.
110. Luvaas, Jay. *Frederick the Great on the Art of War*. New York: The Free Press, 1966.
111. Lynn, John A., ed. *Tools of War: Instruments, Ideas, and Initiations of Warfare, 1445–1871*. Urbana, IL: University of Illinois Press, 1990.
112. MacMunn, Sir George. *Gustavus Adolphus: The Northern Hurricane*. London: Hodder and Stoughton, 1930.
113. Manucy, Albert. *Artillery through the Ages: A Short Illustrated History of Cannon, Emphasizing Types Used in America*. Washington, DC: Government Printing Office, 1949.
114. Mauvillon, J. *Geschichte Ferdinands Herzogs von Braunschweig-Lüneburg*. 2 vols. Leipzig, 1794.
115. May, Sir Edward S. *Achievements of Field Artillery*. Woolwich, England: Royal Artillery Institution, 1893.
116. McNeill, William H. *The Pursuit of Power: Technology, Armed Forces, and Society since A.D. 1000*. Chicago: University of Chicago Press, 1982.
117. Meithem, Michael. *German Treatise on Artillery*. Dresden and Leipzig, 1736.
118. Mirabeau, H. G., and J. Mauvillon. *Système Militaire de Prusse*. London, 1788.
119. Montross, Lynn. *War through the Ages: The Story of War in All Its Phases from 490 B.C.* through World War II and Its Sequel in Korea, Israel, and Egypt to the Present. Rev. ed. New York: Harper and Row, 1960.
120. Moretti, Tomas. *A General Treatise of Artillery*. Translated by Sir Jonas Moore. Brescia, Italy, 1682.
121. Muller, John. *A Treatise on Artillery*. Philadelphia: John Norman, 1759.
122. Norman, Vesey. *Arms and Armour*. New York: G. P. Putnam, 1964.
123. Norman, Vesey, and D. Pottinger. *English Weapons and Warfare: 449–1660*. Englewood Cliffs, NJ: Prentice-Hall, 1979.
124. ———. *Warrior to Soldier*. New York: Crowell, 1966.
125. Norton, Robert. *Gunners Dialogue with the Art of Great Gunnery*. London, 1643.
126. ———. *Of the Art of Great Artillery*. London, 1624.
127. ———. *The Gunner shewing the Whole Practice of Artillery*. London, 1628.

128. Nosworth, Brent. *The Anatomy of Victory: Battle Tactics, 1689–1763*. New York: Hippocrene Books, 1990.
129. Nye, Master Gunner. *Art of Gunnery*. London: William Leak, 1647.
130. O'Connell, Robert L. *Of Arms and Men: A History of War, Weapons, and Aggression*. New York: Oxford University Press, 1989.
131. Oman, Sir Charles. *A History of the Art of War in the Sixteenth Century*. London: Methuen, 1937.
132. ———. *The Art of War in the Middle Ages, 378–1515*. Rev. ed. Ithaca, NY: Cornell University Press, 1960.
133. O'Neill, Bryan. *Castles and Cannon: A Study of Early Artillery Fortifications in England*. Oxford, England: Clarendon Press, 1960.
134. Paret, Peter, ed. *Makers of Modern Strategy*. Princeton, NJ: Princeton University Press, 1986.
135. Parker, Geoffrey. *The Military Revolution: Military Innovation and the Rise of the West, 1500–1800*. New York: Cambridge University Press, 1958.
136. Partington, J. R. *A History of Greek Fire and Gunpowder*. Cambridge, England: W. Heffer and Sons, 1960.
137. Patrick, John M. *Artillery and Warfare during the Thirteenth and Fourteenth Centuries*. Logan, UT: Utah State University Press, 1961.
138. Peterson, Harold L. *Round Shot and Rammers*. Harrisburg, PA: Stackpole Books, 1969.
139. Pratt, Fletcher. "Breitenfeld: The Battle That Changed the World." *Field Artillery Journal*, May 1934, 244–54.
140. ———. "Marignano: The First Artillery Battle." *Field Artillery Journal*, January 1934, 54–61.
141. Quimby, Robert S. *The Background of Napoleonic Warfare: The Theory of Military Tactics in Eighteenth Century France*. New York: AMS Press, 1957.
142. Rathgen, Bernard. *Das Geschütz im Mittelalter*. Berlin: VAI Verlag, 1928.
143. Reid, William. *Arms through the Ages*. New York: Harper and Row, 1976.
144. Ritter, Gerhard. *Frederick the Great: A Historical Profile*. Translated by Peter Paret. London: Eyre and Spottiswoode, 1968.
145. Rivault, Fleurance David de. *Les Élémens de l'Artillerie*. Paris, 1605.
146. Roberts, John. *The Compleat Cannoniere*. London: E. Oakes, 1639.
147. Roberts, Michael. *Gustavus Adolphus: A History of Sweden, 1611–1632*. 2 vols. New York: Longmans, Green, and Company, 1953–1958.
148. Robins, Benjamin. *The New Principles of Gunnery*. London, 1742.
149. Ropp, Theodore. *War in the Modern World*. Rev. ed. New York: Crowell, Collier, and Macmillan, 1965.
150. Rothenberg, Gunther E. *The Art of Warfare in the Age of Napoleon*. Bloomington, IN: Indiana University Press, 1980.
151. Runciman, Steven. *History of the Crusades*. 3 vols. Cambridge, England: Cambridge University Press, 1951–1955.
152. Saint-Rémy, M. Surirey de. *Mémoire d'Artillerie*. Paris: Chez Jean Neaulme, 1741.
153. Sardi, Pietro. *L'Artiglieria*. Venice, 1621.
154. Scharnhorst, Gerhard von. *Über die Wirkung des Feuergewehrs*. Berlin, 1813.
155. Schmidtchen, V. *Bombarden, Befestigungen, Büchsenmeister, von den ersten Mauerbrechern des Spätmittelalters zur Belagerungsartillerie der Renaissance: eine Studie zur Entwicklung der Militärtechnik*. Düsseldorf, 1977.

156. Schneider, Rudolf. "Die Artillerie des Mittelalters," *Neue Jahrbücher für das klassische Altertum*, 1910.
157. Schubert, H. R. "The First Cast-Iron Cannon Made in England." *Journal of the Iron and Steel Institute* 146 (1942).
158. ———. *History of British Iron and Steel Industry from c. 450* B.C. *to* A.D. *1775*. New York: Macmillan, 1957.
159. Simienowicz, Casimir. *The Great Art of Artillery*. Frankfurt: Johann David Zunnes, 1676.
160. Simon, Edith. *The Making of Frederick the Great*. Boston: Little, Brown, 1963.
161. Singer, Charles, E. J. Holmyard, A. R. Hall, and Trevor I. Williams, eds. *A History of Technology*. Oxford, England: Clarendon Press, 1956.
162. Smith, Thomas. *The Art of Gunnery*. London, 1643.
163. Spaulding, Oliver, Hoffman Nickerson, and John W. Wright. *Warfare: A Study of Military Methods from the Earliest Times*. Washington, DC: Infantry Journal Press, 1939.
164. Stevens, Phillip H. *Artillery through the Ages*. Secaucus, NJ: The Citadel Press, 1965.
165. Tartglia, Niccolò. *Nova Scientia*. 1537.
166. Taylor, F. L. *The Art of War in Italy: 1494–1529*. Cambridge, England: Cambridge University Press, 1921.
167. Tousard, Louis de. *American Artillerist's Companion*. 2 vols. Philadelphia: C. and A. Conrad and Company, 1809–1813.
168. Trebilcock, Clive. "Spin-off in British Economic History: Armaments and Industry, 1760–1914." *Economic History Review* 22 (1969): 474–80.
169. Uffano, Don Diego. *Treatise on Artillery and Its Use*. Zutphen, Holland, 1613.
170. U.S. Army Field Artillery School. *History of the Development of Field Artillery Materiel*. Fort Sill, OK: The Field Artillery School, 1940.
171. Vale, M.G.A. "New Techniques and Old Ideals: The Impact of Artillery on War and Chivalry at the End of the Hundred Years War." In C. T. Allmand, ed., *War, Literature, and Politics in the Late Middle Ages*. New York: Barnes and Noble Books, 1976.
172. Valturio, Roberto. *De l'Arte Militare*. Verona, Italy, 1483.
173. Van Creveld, Martin. *Technology and War: From 2000* B.C. *to the Present*. New York: The Free Press, 1988.
174. Verbruggen, J. F. *The Art of Warfare in Western Europe during the Middle Ages*. Amsterdam: North-Holland, 1977.
175. Vogel, H. *Kurzer Bericht der Artillerie-Wissenschaft*. Zurich, 1739.
176. Webb, Henry J. *Elizabethan Military Science: The Books and Practice*. Madison, WI: University of Wisconsin Press, 1965.
177. Weigley, Russell F. *The Age of Battles: The Quest for Decisive Warfare from Breitenfeld to Waterloo*. Bloomington, IN: Indiana University Press, 1991.
178. Weller, Jac. *Weapons and Tactics: Hastings to Berlin*. New York: St. Martin's Press, 1966.
179. Wertime, Theodore. *The Coming of the Age of Steel*. 2 vols. Chicago: University of Chicago Press, 1962.
180. Wilson, A. W. *The Story of the Gun*. Woolwich, England: Royal Artillery Institution, 1944.

181. Wintringham, T. *Weapons and Tactics*. Harmondsworth, England: Penguin Books, 1974.

182. Wise, Sydney F., and Richard A. Preston. *Men in Arms: A History of Warfare and Its Interrelationships with Western Society*. New York: Holt, Rinehart, and Winston, 1979.

183. Wise, Terence. *Medieval Warfare*. New York: Hastings House, 1976.

184. Zook, D. H. and Robin Higham. *A Short History of Warfare*. New York: Twayne, 1965.

6

From Gribeauval to Rifled Field Artillery, 1764–1865

GENERAL HISTORIES

Before examining works on field artillery for the years between 1764 and 1865, studies on warfare covering the years under consideration should be read in addition to the general works cited in the first bibliographical essay. Larry H. Addington's *Patterns of War since the Eighteenth Century* (1) is an excellent place to start. Addington provides a solid synthesis of the changes in warfare from the eighteenth century through the late twentieth century. He examines strategy, technology, military organization, and operations. William McElwee's *The Art of Warfare: Waterloo to Mons* (146) covers a shorter period of time and approaches warfare from a different perspective. After discussing the legacy of Napoleon, he explores war as a means of imperial expansion, the impact of technology, including field artillery, the American Civil War, and other significant topics. Cyril B. Falls's *A Hundred Years of War: 1850–1950* (66) addresses the impact of industrialization on warfare. Published later, Falls's *The Art of War from the Age of Napoleon to the Present Day* (67) also deserves attention for the study of warfare in the nineteenth century. Covering the same years, J.F.C. Fuller's *The Conduct of War: 1789–1961* (75) focuses on the impact of the French, Industrial, and Russian revolutions on war and should be read closely. Hew Strachan's *European Armies and the Conduct of War* (189) is especially valuable for its coverage of tactics, including field artillery. He writes about the eighteenth-century revolution in tactics, Napoleonic warfare, and impact of technology, among other subjects.

Any reading on Napoleonic warfare must begin with David G. Chandler's *The Campaigns of Napoleon* (33). It is the classic study of Napoleon's campaigns. Shorter than Chandler's work but still valuable for anyone's library, Gunther E.

Rothenberg's *The Art of Warfare in the Age of Napoleon* (179) covers tactics, including field artillery, technology, staff work, medical services, and organization. His fine bibliography identifies key studies on Napoleonic warfare. Robert S. Quimby's *The Background of Napoleonic Warfare: The Theory of Military Tactics in Eighteenth Century France* (1973) is critical for understanding field artillery organization, tactics, and equipment and the leading tacticians of the century that influenced Napoleon. Although it is more narrow in scope than Quimby's study, John A. Lynn's *The Bayonets of the Republic: Motivation and Tactics of the Army of Revolutionary France, 1791–1794* (138) is an outstanding book on the French army. He has a particularly informative chapter on cavalry and field artillery. Godfrey Davies's *Wellington and His Army* (49) emphasizes the British experience.

Following the Napoleonic Wars, Baron de Jomini published *The Art of War* (124). He stresses geographic points, speed, movement, lines of supply, and communication as means for success on the battlefield because of his revulsion against the unrestrained warfare of the time. Another contemporary of Napoleon, Carl von Clausewitz, provides in his *On War* (34) an explanation of Napoleon's style of war and a penetrating analysis of the relationship between war and policy. Both of these works are critical for the student of military history.

For those interested in warfare in the United States during the years when the Napoleonic Wars were raging in Europe, John R. Alden's *The American Revolution: 1775–1783* (3) is a basic study and a starting place. Also, Don R. Higginbotham's *The War of American Independence: Military Attitudes, Policies, and Practice, 1763–1789* (106) should be read for its perceptive analysis of the military aspects of the American Revolution; it is one of the best single volumes on the war. Willard M. Wallace's *Appeal to Arms: A Military History of the American Revolution* (202) is especially valuable for infantry and artillery tactics. In his two-volume work *The War of the Revolution* (209), edited by John R. Alden, Christopher Ward provides a solid analysis of the American Revolution at the tactical level. For the next American war consult John K. Mahon's *The War of 1812* (39). Besides furnishing a sound examination of the strategic setting of the war, Mahon explores tactics and operations. Harry L. Coles's *The War of 1812* (37) is shorter than Mahon's fine study but furnishes a constructive introduction to the war, while Donald R. Hickey's *The War of 1812: Forgotten Conflict* (101) covers the military, diplomatic, and domestic history of the war to supply a comprehensive account.

Books on other wars between 1815 and 1865 deserve attention. The standard work on the United States's war with Mexico in 1846–1848 is Justin H. Smith's two-volume *The War with Mexico* (185). K. Jack Bauer's *The Mexican War: 1846–1848* (12) is undoubtedly the best single-volume work on the Mexican War. For the Crimean War see John S. Curtiss's *Russia's Crimean War* (45) and *The Russian Army under Nicholas I, 1825–1855* (46) and R.L.V. Ffrench-Blakes's *The Crimean War* (73). Cecil Woodham-Smith's eloquent *The Reason Why* (217) is another excellent study on the Crimean War.

Although many works on the American Civil War of 1861–1865 are available, only a few are noted here. The most comprehensive general account is Bruce Catton's three-volume *The Centennial History of the Civil War* (31). Another general but worthwhile study is Shelby Foote's *The Civil War: A Narrative History* (72). One of the best single volumes on the Civil War is James M. McPherson's *Battle Cry of Freedom: The Civil War Era* (148). It covers the military as well as the social, economic, and political aspects of the war in an outstanding fashion. For a fine analysis of Civil War tactics and technology see Edward Hagerman's *The American Civil War and the Origins of Modern Warfare: Ideas, Organization, and Field Command* (89).

FIELD ARTILLERY HISTORIES

Several studies address the emergence of true field artillery during the latter years of the eighteenth century. In *Round Shot and Rammers* (169) Harold Peterson supplies line drawings illustrating French, British, and American field artillery systems of the late eighteenth century through the mid-nineteenth century to accompany a fine narrative and a solid bibliography of primary and secondary sources. Albert Manucy's *Artillery through the Ages: A Short-Illustrated History of Cannon, Emphasizing Types Used in America* (142) is another book to consider. Although he focuses on American artillery, Manucy examines the major field artillery systems in Europe during the Napoleonic era. B. P. Hughes's *Firepower: Weapons Effectiveness on the Battlefield: 1630–1850* (113) evaluates the effectiveness of smoothbore field artillery and infantry weapons on the battlefields of Europe and is the standard work on the subject of weapon effectiveness for the years under consideration. Also of value is Philip Haythornthwaite's *Weapons and Equipment of the Napoleonic Wars* (92). In his well-illustrated work the author surveys the artillery, equipment, projectiles, gun crew operations, and national patterns of field artillery of various European armies. Although Sir Edward S. May evaluates field artillery developments through the 1890s, his *Achievements of Field Artillery* (143) is essential for its coverage of field artillery operations, organization, and tactics during the Napoleonic years. In his short book *Artillery: The Big Guns Go to War* (123) Curt Johnson analyzes field artillery developments beginning with the Crimean War through Vietnam. The author supplies photographs and charts that are extremely useful. An old but valuable work on European artillery is C. Beckerhinn's *Die Feld-Artillerie Österreichs, Deutschlands, Englands, Russlands, Italiens und Frankreichs in Bezug auf ihre Bewaffnung, Ausrüstung, Organisation und Leistungsfähigkeit* (The Field Artillery of Austria, Germany, England, Russia, Italy, and France in Reference to their Armament, Equipment, Organization, and Efficiency, 15). His book is especially critical for its discussion of the conversion to rifled steel breechloaders during the latter years of the nineteenth century. Although it does not devote much space to field artillery, Merritt Roe Smith's *Harpers Ferry Armory and the New Technology: The Challenge of Change* (186) addresses the difficulties of adapting to the new rifled technology and provides a background to the rationale for developing rifled field artillery.

An old but worthwhile book is Archibald F. Becke's *An Introduction to the History of Tactics: 1740–1905* (14). Becke scrutinizes infantry and field artillery tactics over the course of almost two centuries. Although it is a general work on artillery and small arms, W. W. Greener's *Gunnery in 1858: Being a Treatise on Rifles, Cannon, and Sporting Arms, etc.* (82) has a small section on field artillery as it existed during the mid-nineteenth century.

Those interested in the field artillery of specific armies should check Christopher Bellamy's *Red God of War: Soviet Artillery and Rocket Forces* (16). Although he concentrates on Soviet artillery, he furnishes an outstanding discussion on Russian field artillery during the Napoleonic period and provides an extensive bibliography of Russian- and English-language books and articles. For a solid examination of Prussian field artillery consult Dennis E. Showalter's *Railroads and Rifles: Soldiers, Technology, and the Unification of Germany* (183A).

Patrick Griffith's *French Artillery* (85) is a superb examination of French field artillery from 1800 through 1815. In his brief illustrated work Griffith discusses the Gribeauval system, weapon capabilities, battery organization and drills, and equipment. Another useful book on French field artillery is James E. Hicks's *French Military Weapons: 1717–1938* (105). Besides covering all French weapons since the early eighteenth century, Hicks writes about the various artillery systems and includes line drawings of several fieldpieces and a table of data on capabilities. The same author's *Notes on French Ordnance: 1717–1936* (102) is also valuable, as is R. Wilkinson-Latham's *Napoleon's Artillery* (213). In *French Napoleonic Artillery* (99) Michael G. Head provides a well-illustrated book on artillery material and uniforms that is worth examining. For a superb reference work consult Philip Haythornthwaite's *The Napoleonic Source Book* (93). The author covers virtually every topic imaginable concerning Napoleon in an encyclopedic fashion but also includes a sound section on field artillery.

An old but still valuable study that covers French artillery is Karl Decker's *The Three Arms* (51). Archibald F. Becke's *Napoleon and Waterloo* (13) offers essential insights into field artillery action during that great battle.

For a general book on French field artillery read General Susane's *Histoire de l'Artillerie Française* (History of French Artillery, 190). It is old but worth examining. Another old study is Jean Baptiste Brunet's *Histoire Générale de l'Artillerie* (A General History of the Artillery, 27). Matti Lauerma's *L'Artillerie de Campagne Française pendant les Guerres de la Révolution* (French Field Artillery during the Wars of the Revolution, 134) examines the changes in French field artillery during the revolution and is the standard work on the subject in French. Although it is old, Jules Challéat's *Histoire Technique de l'Artillerie de Terre en France: 1816–1919* (A Technical History of French Artillery: 1816–1919, 32) is still one of the best works on French field artillery of the nineteenth century. Léon Mention's *L'Armée de l'ancien Régime de Louis XIV à la Rév-*

olution (The Army of the Old Regime of Louis XIV to the Revolution, 151) is also a competent source. De Scheel's *Mémoire d'Artillerie* (Treatise on Artillery, 53) and Saint Auban's *Mémoire sur le Nouveau Système d'Artillerie* (Treatise on the New System of Artillery, 10) are also expert contemporary sources. J. J. Gassendi's *Aide-Mémoire à l'Usage des Officiers d'Artilleries en France* (A Précis on French Artillery for the Use of Officers, 76) should be considered.

For more on French artillery consult Ildefonse Favé's *Nouveau Système d'Artillerie de Campagne de Louis-Napoléon Bonapart* (The Emperor Napoleon's New System of Field Artillery, 70), translated by William H. Cox in 1854. It is a valuable work for French field artillery during the middle of the nineteenth century. The same author's *Histoire de l'Artillerie* (History of Artillery, 68) is another vital study of French field artillery during the early years of the nineteenth century that is worthy of consideration, as is his *Histoire et Tactique des Trois Armes* (The History and Tactics of the Three Arms, 69). M. Favé and Louis Napoleon Bonaparte's (Napoleon III's) six-volume *Études sur le Passé et l'Avenir de l'Artillerie* (Studies on the Past and Future of the Artillery, 71) should also be consulted. In volume six they cover the introduction of rifled artillery in the French army.

For British field artillery see Kenneth Brookes's *Battle Thunder: The Story of Britain's Artillery* (25). It is a general account of British artillery and covers the introduction of true field artillery well. In a well-illustrated book, *British Smooth-Bore Artillery: The Muzzle Loading Artillery of the 18th and 19th Centuries* (112), B. P. Hughes, a gunner himself, covers casting of ordnance, propellants, projectiles, equipment, and specialized forms of artillery, such as mountain and coast artillery. Besides this, he discusses the field artillery of other European armies of the time. Because of its outstanding examination of the topic, Hughes's work is a must for students and scholars to read. In *Open Fire: Artillery Tactics from Marlborough to Wellington* (114), Hughes covers the different types of artillery, fire direction, and artillery organization and has written one of the best books on the topic.

H.W.L. Hime's two-volume *History of the Royal Regiment of Artillery* (107) appraises the achievements of British artillery from the time of Henry VIII to the end of the Napoleonic Wars. Robert Wilkinson-Latham's *British Artillery on Land and Sea: 1790–1820* (212) should also be consulted for its perceptive analysis. Julian R. Jocelyn's *The History of the Royal Regiment* (122) scrutinizes that unit during the Crimean War years and its participation in the Indian Mutiny. Francis Duncan's *History of the Royal Regiment of Artillery* (63) is also useful, while C.A.L. Graham's *The Story of the Royal Regiment of Artillery* (80) offers another perspective on the regiment. A more general account of the Royal Artillery Regiment is Sir Charles E. Caldwell and John Headlam's *The History of the Royal Artillery from the Indian Mutiny to the Great War* (29).

John Hagen in *A New Treatise of Gunnery* (88) outlines gunnery as it was evolving in the 1790s. Charles H. Owens's *Principles and Practice of Modern Artillery, Including Artillery Materiel, Gunnery, and Organization and Use of*

Artillery in Warfare (166) provides a sound examination of British field artillery in the nineteenth century and before and has indispensable footnotes.

For a valuable overview of American field artillery see Boyd L. Dastrup's *King of Battle: A Branch History of the U.S. Army's Field Artillery* (48). It covers tactics, organization, doctrine, materiel, and operations, furnishes a solid overview of the development of American field artillery, and has an extensive bibliography of primary and secondary sources. Another outstanding study on American artillery is William E. Birkhimer's *Historical Sketch of the Artillery of the United States* (23), having its focus on organization, tactics, materiel, and training. The lack of a good index, however, makes using the book difficult.

Donald W. Holtz's "Continental Light Artillery" (111) should also be considered when examining American field artillery. See Don Berkbile's "The 2 3/4-inch U.S. Howitzer, 1792–1793" (21) for a discussion of early pack artillery in the U.S. Army. During the early years of the nineteenth century, the U.S. Army introduced its first horse artillery battery. For a discussion on the topic read Harry C. Larter's "Materiel of the First American Light Artillery, 1808–1809" (132) and "On Horses of the U.S. Artillery, Early 1800's" (133). He discusses the evolution of tactics, doctrine, organization, and weapons and has an outstanding bibliography. Stanley Falk's "Artillery for the Land Service: The Development of a System" (65), focuses on the U.S. Army's efforts during the first four decades of the nineteenth century to develop a field artillery system that was comparable to European systems. Lester R. Dillon's *American Artillery in the Mexican War: 1846–1848* (57) is, perhaps, the standard work on the U.S. Army's field artillery in the Mexican War. In his book he covers tactics, organization, weapons, and operations, and has an invaluable bibliography. Robert Anderson's *An Artillery Officer in the Mexican War: 1846–1848* (7) is indispensable for its description of field artillery action, especially American.

The American Civil War provided armies with one of the first opportunities to test rifled field artillery and feel the impact of rifled muskets on the battlefield. Those interested in an excellent examination of tactics, including field artillery, should consult Paddy Griffith's *Battle Tactics of the Civil War* (84). Grady McWhiney and Perry D. Jamieson's *Attack and Die: Civil War Military Tactics and the Southern Heritage* (149) is worthwhile for its discussion of American field artillery tactics in the Mexican War and the Civil War. For a general account of Civil War field artillery consult Dean S. Thomas's *Cannons: An Introduction to Civil War Artillery* (194) with its excellent drawings of various cannon. Those wanting a lively account of Civil War combat action should read Fairfax Downey's *Sound of the Guns: The Story of American Artillery* (60). Although it is a general history of American field artillery, Downey's descriptions of Civil War battle is useful. Downey's *The Guns at Gettysburg* (61) and *Cannonade: Great Artillery Actions of History, the Famous Cannons, and the Master Gunners* (59) are also engrossing reading. A short article on Civil War field artillery is Jac Weller's "The Field Artillery of the Civil War" (210). A typescript manuscript located in the Morris Swett Technical Library, Fort Sill, Oklahoma, "Remarks upon the Organization, Command and Employment of Field Artillery during War; Based on Experiences of the Civil War, 1861–65" (196), by John C. Tidball, is an outstanding work on Civil War field artillery.

In *The Long Arm of Lee* (216) Jennings C. Wise explores the contributions of General Robert E. Lee's field artillery. Wise eloquently describes tactics, organization, equipment, personalities, and operations. Edward P. Alexander's (q.v.) "Confederate Artillery Service" (4) should also be considered when examining Civil War field artillery, especially Confederate. George M. Neese's *Three Years in the Confederate Horse Artillery* (164) has useful information. For more on Confederate field artillery read Larry J. Daniels's *Cannoneers in Gray: The Field Artillery of the Army of Tennessee* (47). The author explains clearly the difficulties encountered in the western campaigns of the Civil War and the challenges in adapting to the new rifled technology.

The definitive account of the field artillery in the Army of the Potomac is L. Van Loan Naisawald's *Grape and Canister: The Story of the Field Artillery of the Army of the Potomac, 1861–1865* (162). The author discusses organization, tactics, equipment, weapons, operations, and important personalities. For another look at Union artillery consult J. G. Barnard and W. F. Barry's *Report of the Engineer and Artillery Operations of the Army of the Potomac from Its Organization to the Close of the Peninsular Campaign* (11). In *The Rhode Island Artillery at the First Battle of Bull Run* (153) John A. Monroe discusses the failure of Napoleonic-style smoothbore artillery tactics with the appearance of rifled muskets.

In the 1960s James C. Hazlett wrote "The Confederate Napoleon Gun" (95), "The Federal Napoleon Gun" (96), "The Napoleon Gun: Its Origin and Introduction into American Service" (97), and "The Napoleon Gun: Markings, Bore Diameters, Weights, and Costs" (98). All are worthy of reading and serious consideration when examining American smoothbore artillery during the middle of the nineteenth century.

Battles and Leaders of the Civil War, Vol. III, contains useful articles on field artillery during the American Civil War. See Henry J. Hunt's "The First Day at Gettysburg" (116), "The Second Day at Gettysburg" (117), and "The Third Day at Gettysburg" (118). Other solid articles are Edward P. Alexander's "The Great Charge and Artillery Fighting at Gettysburg" (6), Francis Walker's "General Hancock and the Artillery at Gettysburg" (201), and Henry J. Hunt's "General Hancock and the Artillery at Gettysburg" (115).

Several useful biographies are available on key Civil War personalities. *The Story of a Cannoneer under Stonewall Jackson* (154) by Edward A. Moore provides insights into Confederate field artillery during the American Civil War. Philip Mercer's *The Life of Gallant Pelham* (152) covers the activities of a famous Confederate field artilleryman, John Pelham, who was noted for his daring employment of field artillery. John W. Morton's *Forrest's Artillery* (160) is worth reading about Nathan Bedford Forrest's field artillery. Douglas S.

Freeman's three-volume *Lee's Lieutenants: A Study in Command* (74) contains portraits of a number of Confederate artillerymen. Edward P. Alexander's *Military Memoirs of a Confederate: A Critical Narrative* (5) should be examined. Edward G. Longacre's *The Man behind the Gun: A Biography of General Henry Jackson Hunt, Chief of Artillery, Army of the Potomac* (136) is the definitive account of Hunt's contributions during the American Civil War as well as of his life as a whole. For an autobiography of a Civil War artilleryman see August Buell's *The Cannoneer: Recollections of Service in the Army of the Potomac* (28).

Other personalities played a critical role in the development of field artillery. Although it is old, Eugène Hennebart's *Gribeauval* (100) is a starting point for studying the famous French artilleryman. For other notable French artillerymen see de l'Ain Girod's *Grands Artilleurs—Druot, Sernamount-Elbe* (The Great Artillerymen: Druot and Sernamount-Elbe, 78). Noah Brooks's *Henry Knox: A Soldier of the Revolution* (26) is a short account, while North Callahan's *Henry Knox: General Washington's General* (30) is the definitive study of Knox (q.v.), the father of American field artillery. Henry Bessemer's *An Autobiography* (22) is worth consulting for metallurgy. J. Emerson Tennant's *The Story of the Guns* (192) covers Whitworth's contributions to artillery developments as well as the difficulties encountered with early rifled field artillery, while David Dougan's *The Great Gunmaker: The Story of Lord Armstrong* (58) examines Whitworth's major competitor in rifled field artillery.

For works on the Krupps see Wilhelm Berdrow's *The Krupps: 150 Years of Krupp History, 1787–1939* (19), William Manchester's dramatic account, *The Arms of Krupp* (141), and Gert von Klass's *Krupps: The Story of an Industrial Empire* (127). Bernard Menne's *Krupp; or, The Lords of Essen* (150) provides another perspective on the Krupp industry.

During the period under consideration, rockets were being introduced after years of neglect. For a sound discussion see Frank H. Winter's *The First Golden Age of Rocketry: Congreve and Hale Rockets of the 19th Century* (215). In this fine work the author explores the impact of rockets on warfare and the achievements of Congreve and Hale in light of other types of artillery, and has a substantial bibliography of secondary and primary sources.

TECHNOLOGY

Even though it is a general history of artillery, Joseph Jobe's (ed.) *Guns: An Illustrated History of Artillery* (121) covers field artillery technology well. O.F.G. Hogg's *Artillery: Its Origin, Heyday, and Decline* (108) has a particularly useful section on rifled artillery, shrapnel projectiles, and early work with high explosives as a propellant and bursting charge. Although A. W. Wilson's *The Story of the Gun* (214) furnishes a general history of the development of artillery, it also offers insights into the development of rifled artillery in the mid-eighteenth century. Frank E. Comparato's *Age of Great Guns: Cannon Kings and Cannoneers Who Forged the Firepower of Artillery* (39) has vital information on the introduction of rifled field artillery in the 1840s and 1850s and an extensive bibliography. William H. McNeill's *The Pursuit of Power: Technology, Armed Forces, and Society since* A.D. *1000* (147) has an excellent section dealing with the introduction of rifled field artillery.

Contemporary writings of field artillery are available for those interested. For early nineteenth century consult Isaac Landmann's *Principles of Artillery* (131). Written in the mid-nineteenth century, Robert Mallet's *The Construction of Artillery* (140) discusses design and construction of artillery. Thomas Lyndall's *Rifled Ordnance: A Practical Treatise on the Application of the Principle of Rifle to Guns and Mortars of Every Caliber* (137) provides a practical analysis of rifled artillery during the middle decades of the nineteenth century. Richard Delafield's *Report on the Art of War in Europe in 1854, 1855, and 1856* (52) supplies insights into the critical field artillery developments during the 1850s. In *Elementary Lectures on Artillery as Delivered at the Royal Military Academy* (165), Charles H. Owens and T. L. Dames present a comprehensive but elementary survey of the development of artillery through the 1850s as does John Scoffern in *New Resources of Warfare with Especial Reference to Rifled Ordnance in Their Chief Known Varieties* (183). Alexander L. Holley's *A Treatise on Ordnance and Armor* (110) is an excellent account of the state of field artillery and ordnance in the 1860s. Although it is short, John W. Cochran's *Improvements in Ordnance, Firearms, and Projectiles* (35) is worth examining.

W. W. Greener's *The Gun and Its Development* (83) discusses the introduction of rifled artillery. Franck Taylor's *Rifled Field Pieces: A Short Compilation of What Is Known of the New Field Artillery in Europe with Some Accounts of Our Own* (191) is worth reading for its discussion of rifled field artillery in Europe and the United States in the 1850s. Alfred Mordecai's *Military Commission to Europe in 1855 and 1856* (156) analyzes field artillery technology of the 1850s, while Thomas J. Rodman's *Reports of Experiments on the Properties of Metals of Cannons and the Qualities of Cannon Powder with an Account of the Fabrication and Trials of a 15-inch Gun* (178) covers cannon made by Rodman. For a lengthy but solid study consult L. F. Mazé's seven-volume *État Actuel de l'Artillerie de Campagne en Europe* (The Present State of Field Artillery in Europe, 144), J. P. Schenkl's *Description of the Combination and Concussion Fuze* (182) explains the development of percussion fuzes during the mid-nineteenth century.

For rocket artillery during the late eighteenth and early nineteenth centuries consult William Congreve's *A Treatise on the General Principles, Powers and Facility of Application of the Congreve Rocket System as Compared with Artillery* (41). It is invaluable for its discussion of the application of rocketry to sea and land warfare. Also see Congreve's *A Concise Account of the Origin and Progress of the Rocket System* (40). William Hale's *The Comparative Merits of a Rifle*

Gun and a Rotary Rocket (91) and *Hale's War Rockets, Statements for the Referee* (90) are also valuable for their discussions on field artillery materiel at mid-century and its comparison with rockets.

For British and American artillery developments of the late eighteenth century see John Muller's *A Treatise of Artillery* (161), which is the standard work. He outlines carriage, limber, and caisson patterns and casting instructions. Edward M. Boxer's *Treatise on Artillery* (24) studies British field artillery through the Crimean War and includes scale drawings of all the pieces in the British service at the time. J. Pottinger's *A Plain Treatise on Artillery* (170) supplies valuable line drawings of ammunition and carriages of the Bombay Artillery in the British army during the early years of the nineteenth century. Another book on British technology is O.F.G. Hogg's *The Royal Arsenal: Its Background, Origin, and Subsequent History* (109).

Jean-Baptiste Gribeauval's *Tables des Constructions des Principaux Attirails de l'Artillerie* (Tables of Construction for Artillery Equipment, 87) is critical for studying French artillery. L. F. Mazé's *Artillerie de Campagne en France, Description de l'Organisation et du Matériel de cette Arme en 1845* (French Field Artillery: The Description of the Organization and Equipment of the Arm in 1845, 145) is also worth perusing for organization and materiel.

For studies written in the twentieth century on American field artillery read James E. Hicks's *Notes on U.S. Ordnance, 1776–1941* (104). It provides a more general approach to American ordnance during the Civil War era. James C. Hazlett's *Field Artillery Weapons of the Civil War* (94) covers field artillery in the Union and Confederate armies during the American Civil War. The best work on Civil War artillery is Warren Ripley's *Artillery and Ammunition of the Civil War* (174). In this comprehensive study he examines the history of various field artillery pieces and ammunition, includes pictures, and explains the technology. Jack Coggin's *Arms and Equipment of the Civil War* (36) is useful for its information of rifled artillery. Other works on Civil War artillery include: Thomas S. Dickey and Sidney Kerkis's *Heavy Artillery Projectiles of the Civil War, 1861–1865* (56) and Thomas S. Dickey and Peter C. George's *Field Artillery Projectiles of the Civil War* (55).

Contemporary works are abundant for American field artillery technology. See M. De Scheel's *A Treatise of Artillery Containing a New System or the Alterations Made in the French Artillery since 1765* (54), translated by Jonathan Williams. It is necessary when examining American field artillery developments during the first ten years of the nineteenth century. De Scheel outlines the Gribeauval system, which was modified for American use. Louis de Tousard's *American Artillerist's Companion* (197) is a standard work on ordnance for the first decade of the nineteenth century. In his two-volume study he covers the Gribeauval reforms, gives particulars on performance, and provides useful scale drawings of equipment and carriages.

Of special interest is Alfred Mordecai's (q.v.) *Artillery for the Land Service of the United States* (155). It is especially vital for its drawings of the bronze smoothbore field artillery that formed the basis of American field artillery materiel at mid-century. Mordecai's *Reports on Experiments with Gunpowder at Washington Arsenal in 1843 and 1844* (157) and *Second Report on Experiments on Gunpowder Made at Washington Arsenal in 1845, 1847, 1848* (158) and *The Ordnance Manual for the Use of Officers of the United States Army*l (159), are also worth examining. Although the latter is short, it contains much priceless information. J. N. Paulding's *The Cannon and Projectiles Invented by Robert Parker Parrott* (168) is valuable for early American rifled field artillery technology. J. G. Benton's *A Course of Instruction in Ordnance and Gunnery* (18) is also worth consulting. Like many other writers of the times, Benton covers gunpowder, projectiles, cannon carriages, gunnery, and employment of artillery. Daniel Treadwell's "On the Construction of Hooped Cannon, Being a Sequel to a Memoir on the Practicability of Constructing Cannon of Great Caliber" (198) is invaluable for the development of hooped rifled artillery.

Additional works on American developments exist. Less well-known are Joseph Roberts's *The Hand-book of Artillery of the Service of the United States* (175), *The Handbook of Artillery* (176), and *The Handbook of Artillery for the Service of the U.S. Army and Militia* (177). Alexander L. Holley's *A Treatise on Ordnance and Armor* (110) is also of merit. For a discussion of rifled field artillery consult *Wiard's System of Field Artillery* (211) by Norman Wiard. The U.S. Ordnance Department's *The Ordnance Manual for the Use of Officers of the United States Army* (199) and *The Ordnance Manual for the Use of Officers of the United Stats Army* (200), the latter published in Philadelphia, are also invaluable for their discussion of field artillery materiel, ordnance, and munitions. Henry Jervis-White's *Our Engines of War and How We Got to Make Them* (120) has solid coverage of rifled artillery. U.S. Ordnance Department's four-volume *Collection of Reports: 1812–1889 (17),* edited by Stephen Vincent Benét (q.v.), is an invaluable collection of primary source documents on the development of American ordnance during the nineteenth century. In *Report on the Munitions of War* (163) Charles B. Norton discusses American field artillery munitions of the 1860s, while Josiah Gorgas examines Confederate ordnance in *Ordnance Field Manual* (79).

For German artillery technology see James Hicks's *Notes on German Ordnance: 1841–1918* (103), which has a solid section on field artillery with accompanying line drawings and characteristics of weapons. An indispensable study on German field artillery is H. F. Kameke's *Sammlung von Zeichnungen, die Einrichtung der materiellen Gegenstände der preussischen Artillerie darstellend, und Eräuterungen dazu* (A Collection of Designs for the Organization of Prussian Artillery Materiel, 125), while F. Berger's *Zeichnungen des königlichen preussischen Artillerie Materiels* (Designs of the Royal Prussian Artillery Materiel, 20) details Prussian field artillery at mid-century.

TACTICS, DOCTRINE, AND ORGANIZATION

Between 1765 and 1865, numerous drill and technical manuals appeared that outlined gun drill, the means for maneuvering batteries on the battlefield, and procedures for casting ordnance and making equipment. Perhaps one of the most essential works on French field artillery of the Napoleonic era is Jean Du Teil's *De l'Usage de l'Artillerie Nouvelle dans la Guerre de Campagne* (On the Use of the New Artillery, 64) because of its impact upon French field artillery tactics and organization during the latter years of the eighteenth century and the early years of the nineteenth century. Tronson du Coudray's *L'Artillerie Nouvelle* (The New Artillery, 43) is another solid source on French field artillery drill during the latter years of the eighteenth century. Charles Hutton's *Nouvelles Expériences d'Artillerie* (New Experiences of the Artillery, 119) also explores French field artillery.

For French field artillery following the Napoleonic Wars see M. Poumet's *Analyse et Dissertation Relative à la Nouvelle Artillerie* (Précis and Treatise on the New Artillery, 171). G. H. Cotty's *Dictionnaire de l'Artillerie* (Dictionary on Artillery, 42) and P. A. Poumet's *Instruction sur l'Artillerie de Campagne* (Instruction on Field Artillery, 172) are solid manuals. Espiard de Cologne's *Artillerie Pratique* (The Application of Artillery, 38) is vital. Other works on French artillery are Laguionie Thiroux's *Instruction Théorique et Pratique d'Artillerie* (Theoretical Instruction and Application for the Artillery, 193), J. H. LeBourg's *Essai sur l'Organisation de l'Artillerie et son Emploi la Guerre de Campagne* (Essay on the Organization of the Artillery and Its Employment in War, 135), E. Decker's *Traité de l'Art Combattre de l'Artillerie à Cheval Réunir à la Cavalrie* (A Treatise on Horse Artillery with the Cavalry in Combat, 50), and L. de Crepy's *Cours de'Instruction Speciale à l'Usage des Sous-Officiers des Régiments d'Artillerie* (A Special Course of Instruction for the Use of Artillery Officers, 44).

For American artillery see William Stevens's *A System for the Discipline of the Artillery of the United States of America* (188). In three volumes Stevens covers the maneuvers of horse artillery and artillery construction. Another critical work is Thaddeus Kosciusko's (q.v.) *Maneuvers of Horse Artillery* (129). The author provides one of the first manuals on horse artillery maneuvers and organization in the U.S. Army. Another Kosciusko book is *Exercises for Garrison and Field Ordnance* (128). William Duane's *A Military Dictionary* (62) examines discipline and drill for infantry, artillery, and cavalry. Another book written by an American is Amasa Smith's *A Short Compendium of the Duty of Artillerists* (184). Henri D. Lallemand's *Treatise on Artillery* (130), translated by James Renwick, discusses maneuvers of horse artillery as well as organization, and was used extensively by the U.S. Army. The U.S. War Department's *A System of Exercise and Instruction of Field Artillery Including Maneuvers for Light or Horse Artillery* (203) is a useful manual. For more about American horse artillery examine Robert Anderson's *Instruction for Field Artillery, Horse and Foot* (8), a translation from a French manual. Its coverage of organization, gun drill, and maneuver make it indispensable for understanding the U.S. Army's field artillery during the 1840s.

For American field artillery drill during the middle decades of the nineteenth century consult the U.S. War Department's *Evolutions of Field Batteries of Artillery* (204), translated from the French by Robert Anderson. Also see the War Department's *Instruction for Field Artillery* (205), written by William H. French, William F. Barry (q.v.), and Henry J. Hunt (q.v.). The manual went through several editions in the 1860s and formed the basis of Union field artillery during the American Civil War. The War Department's *Instruction for Field Artillery* (206) is also valuable. Another mid-century artillery drill manual in the United States is John Gibbon's *The Artillerist's Manual* (77). Gibbon's manual deals with many of the last smoothbore field artillery pieces before they became obsolete. Still another drill manual is George Patten's *Artillery Drill* (167). The War Department's *Instruction for Field Artillery: Horse and Foot* (1845 edition, 207) and *Instruction for Field Artillery: Horse and Foot* (1850 edition, 208) are invaluable for their discussion of gun drill and organization. Less well-known is C. P. Kingsbury's *An Elementary Treatise on Artillery and Infantry, Adapted to the Service of the United States* (126). A. W. Stark's *Instruction for Field Artillery* (187) is solid for Confederate field artillery, even though it was published late in the Civil War. Robert Anderson's *Evolution of Field Batteries of Artillery* (9) is also worth examining.

During the nineteenth century, the British and Prussians were also actively writing drill manuals. Ralph W. Adye's *The Bombardier and the Pocket Gunner* (2) is useful. Between 1800 and 1827 the book went through eight editions. It provides a wealth of information about artillery equipment for the period. The manual shows dates when new equipment was introduced, weights, dimensions, and performance of individual weapons. Published later, Frederick A. Griffiths's *The Artillerist's Manual and British Soldier's Compendium* (86) furnishes extensive information on the dimensions and performance of British artillery through the Crimean War.

Written by the Prussian general Gerhard von Scharnhorst, *Handbuch der Artillerie* (The Manual of Artillery, 181) covers Prussian artillery drill and tactics. F. G. Rouvroy's *Vorlesungen über die Artillerie* (Lectures on Artillery, 180) also covers Prussian field artillery, as does W. von Grawenitz's *Organisation und Taktik der Artillerie* (Organization and Tactics of Artillery, 81).

BIBLIOGRAPHY

1. Addington, Larry H. *Patterns of War since the Eighteenth Century*. Bloomington, IN: Indiana University Press, 1984.
2. Adye, Robert W. *The Bombardier and the Pocket Gunner*. Boston: E. Larkin, 1804.
3. Alden, John R. *The American Revolution: 1775–1783*. New York: Harper and Row, 1954.
4. Alexander, Edward P. "Confederate Artillery Service." *Southern Historical Society Papers* 11 (1883): 98–113.
5. ———. *Military Memoirs of a Confederate: A Critical Narrative*. New York: Charles Scribner's Sons, 1907.
6. ———. "The Great Charge and Artillery Fighting at Gettysburg." In R. U. Johnson

and C. C. Buel, eds., *Battles and Leaders of the Civil War*. New York: The Century Company, 1887–1888. Reprint. Secaucus, NJ: Castle, 1985.

7. Anderson, Robert. *An Artillery Officer in the Mexican War: 1846–1848*. Philadelphia: Robert P. DeSilver, 1911.
8. ———. *Instruction for Field Artillery, Horse and Foot*. Philadelphia: Robert P. DeSilver, 1839.
9. ———. *Evolutions of Field Batteries of Artillery*. New York, 1860.
10. Auban, Saint. *Mémoire sur le Nouveau Système d'Artillerie*. Paris, 1775.
11. Barnard, John G., and William F. Barry. *Report of the Engineer and Artillery Operations of the Army of the Potomac from Its Organization to the Close of the Peninsular Campaign*. New York: D. Van Nostrand, 1863.
12. Bauer, K. Jack. *The Mexican War: 1846–1848*. New York: Macmillan, 1974.
13. Becke, Archibald. *Napoleon and Waterloo*. 2 vols. London: Kegan Paul, Trench, Truebner, and Company, 1914.
14. ———. *An Introduction to the History of Tactics: 1740–1905*. London: Hugh Rees, 1909.
15. Beckerhinn, C. *Die Feld-Artillerie Österreichs, Deutschlands, Englands, Russlands, Italiens und Frankreichs in Bezug auf ihre Bewaffnung, Ausrüstrung, Organisation und Leistungsfähigkeit*. Vienna, 1879.
16. Bellamy, Christopher. *Red God of War: Soviet Artillery and Rocket Forces*. Washington, DC: Brassey's Defence Publishers, 1986.
17. Benét, Stephen Vincent, ed. *Collection of Reports: 1812–1889*. Washington, DC: Government Printing Office, 1812–1889.
18. Benton, J. G. *A Course of Instruction in Ordnance and Gunnery*. 2nd ed. New York: D. Van Nostrand, 1862.
19. Berdrow, Wilhelm. *The Krupps: 150 Years of Krupp History, 1787–1937*. Berlin, 1937.
20. Berger, F. *Zeichnungen des Königlichen Preussischen Artillerie Materiels*. Berlin, 1856–1857.
21. Berkbile, Don. "The 2 3/4-inch U.S. Howitzer, 1792–1793." *Military Collector and Historian*, Spring 1961.
22. Bessemer, Sir Henry. *An Autobiography*. London, 1905.
23. Birkhimer, William E. *Historical Sketch of the Artillery of the United States*. Washington, DC: John Chapman, 1884. Reprint ed., Westport, CT: Greenwood Press, 1968.
24. Boxer, Edward M. *Treatise on Artillery*. London: George E. Eyre, 1853.
25. Brookes, Kenneth. *Battle Thunder: The Story of Britain's Artillery*. Reading, England: Osprey Publishing Ltd., 1973.
26. Brooks, Noah. *Henry Knox: A Soldier of the Revolution*. New York: DeCapo Press, 1899.
27. Brunet, Jean Baptiste. *Histoire Générale de l'Artillerie*. 2 vols. Paris: Gaultier-Laguionie, Maison Anselin, 1842.
28. Buell, Augustus. *The Cannoneer: Recollections of Service in the Army of the Potomac*. Washington, DC: The National Tribune, 1890.
29. Caldwell, Sir Charles E., and John Headlam. *The History of the Royal Artillery from the Indian Mutiny to the Great War*. Woolwich, England: Royal Artillery Institution, 1931.

30. Callahan, North. *Henry Knox: General Washington's General*. New York: A. S. Barnes and Company, 1958.
31. Catton, Bruce. *The Centennial History of the Civil War*. 3 vols. Garden City, NY: Doubleday, 1961–1965.
32. Challéat, Jules M. J. *Histoire Technique de l'Artillerie de Terre en France: 1816–1919*. Paris: Imprimerie Nationale, 1933.
33. Chandler, David G. *The Campaigns of Napoleon*. New York: Macmillan, 1966.
34. Clausewitz, Carl von. *On War*. Translated by Peter Paret and Michael Howard. Princeton, NJ: Princeton University Press, 1976.
35. Cochran, John W. *Improvements in Ordnance, Firearms, and Projectiles*. New York: J. W. Orr, 1860.
36. Coggin, Jack. *Arms and Equipment of the Civil War*. Garden City, NY: Doubleday, 1962.
37. Coles, Harry L. *The War of 1812*. Chicago: University of Chicago Press, 1965.
38. Cologne, Espiard de. *Artillerie Pratique*. 2 vols. Paris, 1846.
39. Comparato, Frank E. *Age of Great Guns: Cannon Kings and Cannoneers Who Forged the Firepower of Artillery*. Harrisburg, PA: The Stackpole Company, 1965.
40. Congreve, William. *A Concise Account of the Origin and Progress of the Rocket System*. Dublin: A O'Neill, 1817.
41. ———. *A Treatise on the General Principles, Powers, and Facility of Application of the Congreve Rocket System as Compared with Artillery*. London, 1827.
42. Cotty, G. H. *Dictionnaire de l'Artillerie*. Paris, 1822.
43. Courdray, Tronson du. *L'Artillerie Nouvelle*. Amsterdam: Arkster and Merkus, 1772.
44. Crepy, L. de. *Cours de'Instruction Speciale à l'Usage des Sous-Officiers des Régiments d'Artillerie*. Metz: Verronnais, 1837.
45. Curtiss, John S. *Russia's Crimean War*. Ithaca, NY: Cornell University Press, 1979.
46. ———. *The Russian Army under Nicholas I, 1825–1855*. Durham, NC: Duke University Press, 1965.
47. Daniels, Larry J. *Cannoneers in Gray: The Field Artillery of the Army of Tennessee*. University, AL: University of Alabama Press, 1984.
48. Dastrup, Boyd L. *King of Battle: A Branch History of the U.S. Army's Field Artillery*. Fort Monroe, VA: Office of the Command Historian, U.S. Army Training and Doctrine Command, 1992.
49. Davies, Godfrey. *Wellington and His Army*. Oxford, England: Blackwell, 1954.
50. Decker, E. *Traité de l'Art Combattre de l'Artillerie à Cheval Réunir à la Cavalrie*. Paris: F. G. Levrault, 1831.
51. Decker, Karl. *The Three Arms*. London: Parker, Furnisall, and Parker, 1851.
52. Delafield, Richard. *Report on the Art of War in Europe in 1854, 1855, 1856*. Washington, DC: George W. Bowman, 1861.
53. De Scheel, M. *Mémoire d'Artillerie*. Copenhagen, 1777.
54. ———. *A Treatise of Artillery Containing a New System or the Alterations Made in the French Artillery since 1765*. Translated by Jonathan Williams. Philadelphia, 1800.
55. Dickey, Thomas S., and Peter C. George. *Field Artillery Projectiles of the Civil War*. Atlanta: Arsenal Press, 1980.

56. Dickey, Thomas S., and Sidney Kerkis. *Heavy Artillery Projectiles of the Civil War, 1861–1865*. Atlanta: Phoenix Press, 1968.
57. Dillon, Lester R. *American Artillery in the Mexican War: 1846–1847*. Austin, TX: Presidial Press, 1975.
58. Dougan, David. *The Great Gunmaker: The Story of Lord Armstrong*. Newcastle upon Tyne, England, n.d.
59. Downey, Fairfax. *Cannonade: Great Artillery Actions of History, the Famous Cannons, and the Master Gunners*. Garden City, NY: Doubleday, 1966.
60. ———. *Sound of the Guns: The Story of American Artillery*. New York: David McKay, 1955.
61. ———. *The Guns at Gettysburg*. New York: David McKay, 1958.
62. Duane, William. *A Military Dictionary*. Philadelphia, 1810.
63. Duncan, Francis. *History of the Royal Regiment of Artillery*. 2 vols. London: John Murray, 1872.
64. Du Teil, Jean. *De l'Usage de l'Artillerie Nouvelle dans la Guerre de Campagne*. Metz, 1778.
65. Falk, Stanley. "Artillery for the Land Service: The Development of a System." *Military Affairs*, Fall 1964, 97–110.
66. Falls, Cyril B. *A Hundred Years of War: 1850–1950*. New York: Capricorn Press, 1953.
67. ———. *The Art of War from the Age of Napoleon to the Present Day*. New York: Oxford University Press, 1961.
68. Favé, I. *Histoire de l'Artillerie*. London: Parker, Furnisall, and Parker, 1871.
69. ———. *Histoire et Tactique des Trois Armes*. Paris: J. Dumaine, 1851.
70. ———. *The Emperor's New System of Field Artillery*. Translated by William H. Cox. London: Parker, Furnisall, and Parker, 1854.
71. Favé, M., and Louis Napoleon Bonaparte (Napoleon III). *Études sur le Passé et l'Avenir de l'Artillerie*. 6 vols. Paris: J. Dumaine, 1846–1871.
72. Foote, Shelby. *The Civil War: A Narrative History*. 3 vols. New York: Random House, 1958–1974.
73. Ffrench-Blake, R.L.V. *The Crimean War*. Hamden, CT: Archon Books, 1972.
74. Freeman, Douglas. *Lee's Lieutenants: A Study in Command*. 2 vols. New York: Charles Scribern's Sons, 1942–1944.
75. Fuller, J.F.C. *The Conduct of War: 1789–1961*. New Brunswick, NJ: Rutgers University Press, 1961.
76. Gassendi, J. J. *Aide-Mémoire à l'Usage des Officiers d'Artillerie en France*. Paris: Veuve Levrault, 1844.
77. Gibbon, John. *The Artillerist's Manual*. New York: D. Van Nostrand, 1863.
78. Girod, de l'Ain. *Grands Artilleurs—Druot, Sernamount-Elbe*. Paris: Berger-Levrault, 1895.
79. Gorgas, Josiah. *Ordnance Field Manual*. Confederate States of America, 1862.
80. Graham, C.A.L. *The Story of the Royal Regiment of Artillery*. Woolwich, England: Royal Artillery Institution, 1928.
81. Grawenitz, W. von. *Organisation und Taktik der Artillerie*. Paris, 1831.
82. Greener, W. W. *Gunnery in 1858: Being a Treatise on Rifles, Cannon, and Sporting Arms, etc.* 6th ed. New York: Charles Scribner's Sons, 1897.
83. ———. *The Gun and Its Development*. 6th ed. New York: Charles Scribner's Sons, 1897.

84. Griffith, Paddy. *Battle Tactics of the Civil War*. New Haven, CT: Yale University Press, 1987.
85. Griffith, Patrick. *French Artillery*. London: Almark Publishing Company, Ltd., 1976.
86. Griffiths, Frederick A. *The Artillerist's Manual and British Soldier's Compendium*. London: William Clowes and Sons, 1862.
87. Gribeauval, Jean-Baptiste. *Tables des Constructions des Principaux Attirails de l'Artillerie*. Paris, 1792.
88. Hagen, John. *A New Treatise of Gunnery*. Dungannon, Northern Ireland, 1797.
89. Hagerman, Edward. *The American Civil War and the Origins of Modern Warfare: Ideas, Organization, and Field Command*. Bloomington, IN: Indiana University Press, 1988.
90. Hale, William. *Hale's War Rockets, Statements for the Referee*. London, 1865.
91. ———. *The Comparative Merits of a Rifle Gun and a Rotary Rocket*. London: W. Twitchell, 1863.
92. Haythornthwaite, Philip. *Weapons and Equipment of the Napoleonic Wars*. Poole, England: Blandford Press, 1979.
93. ———. *The Napoleonic Source Book*. New York: Facts on File, 1990.
94. Hazlett, James C. *Field Artillery Weapons of the Civil War*. Newark, DE: University of Delaware Press, 1983.
95. ———. "The Confederate Napoleon Gun." *Military Collector and Historian*, Winter 1964, 104–10.
96. ———. "The Federal Napoleon Gun." *Military Collector and Historian*, Winter 1963, 103–8.
97. ———. "The Napoleon Gun: Its Origin and Instruction into American Service." *Military Collector and Historian*, Spring 1963, 1–5.
98. ———. "The Napoleon Gun: Markings, Bore Diameters, Weights, and Costs." *Military Collector and Historian*, Winter 1966, 109–19.
99. Head, Michael G. *French Napoleonic Artillery*. London: Almark Publishing Company Ltd., 1970.
100. Hennebart, Eugène. *Gribeauval*. Paris: Berger-Levrault, 1896.
101. Hickey, Donald R. *The War of 1812: Forgotten Conflict*. Urbana, IL: University of Illinois Press, 1989.
102. Hicks, James E. *Notes on French Ordnance,: 1717–1936*. Mt. Vernon, NY: James E. Hicks, 1938.
103. ———. *Notes on German Ordnance: 1841–1918*. New York: I. Oran, 1937.
104. ———. *Notes on U.S. Ordnance, 1776–1941*. Green Farms, CT: Modern Books and Craft, Inc., 1971.
105. ———. *French Military Weapons: 1717–1938*. Mt. Vernon, NY: James E. Hicks, 1938.
106. Higginbotham, Don R. *The War of American Independence: Military Attitudes, Policies, and Practice, 1763–1789*. New York: Macmillan, 1971.
107. Hime, H.W.L. *History of the Royal Regiment of Artillery*. 2 vols. London: Longmans Green, 1908.
108. Hogg, O.F.G. *Artillery: Its Origin, Heyday, and Decline*. Hamden, CT: Archon Books, 1970.
109. ———. *The Royal Arsenal: Its Background, Origin, and Subsequent History*. London: Oxford University Press, 1963.

110. Holley, Alexander L. *A Treatise on Ordnance and Armor*. New York: D. Van Nostrand, 1865.

111. Holtz, Donald W. "Continental Light Artillery." *Military Collector and Historian*, Summer 1989, 55.

112. Hughes, B. P. *British Smooth-Bore Artillery: The Muzzle Loading Artillery of the 18th and 19th Centuries*. Harrisburg, PA: Stackpole Books, 1969.

113. ———. *Firepower: Weapons Effectiveness on the Battlefield, 1630–1850*. New York: Charles Scribner's Sons, 1974.

114. ———. *Open Fire: Artillery Tactics from Marlborough to Wellington*. Chichester, England: Antony Bird Publications Ltd., 1983.

115. Hunt, Henry J. "General Hancock and the Artillery at Gettysburg." In Robert Underwood and Clarence C. Buel's, eds., *Battles and Leaders of the Civil War*. Vol. III, New York: The Century Co., 1887–1888.

116. ———. "The First Day at Gettysburg." In Robert Underwood and Clarence C. Buel's, eds., *Battles and Leaders of the Civil War*. Vol. III, New York: The Century Co., 1887–1888.

117. ———. "The Second Day at Gettysburg." In Robert Underwood and Clarence C. Buel's, eds., *Battles and Leaders of the Civil War*. Vol. III, New York: The Century Co., 1887–1888.

118. ———. "The Third Day at Gettysburg," In Robert Underwood and Clarence C. Buel's, eds., *Battles and Leaders of the Civil War*. Vol. III, New York: The Century Co., 1887–1888.

119. Hutton, Charles. *Nouvelles Experiences d'Artillerie*. Paris, 1802.

120. Jervis-White, Henry. *Our Engines of War and How We Got to Make Them*. London: Chapman and Hall, 1859.

121. Jobe, Joseph, ed. *Guns: An Illustrated History of Artillery*. New York: New York Graphic Society, 1971.

122. Jocelyn, Julian R. *The History of the Royal Regiment*. London: John Murray, 1911.

123. Johnson, Curt. *Artillery: The Big Guns Go to War*. London: Octopus Books, 1975.

124. Jomini, Baron von. *The Art of War*. Translated by G. H. Mendall and W. P. Craighill. Westport, CT: Greenwood Press, 1974.

125. Kameke, H. F. *Sammlung von Zeichnungen, die Einrichtung der materiellen Gegenstände der preussischen Artillerie darstellend, und Erläuterungen dazu*. Berlin, 1837 and 1843.

126. Kingsbury, C. P. *An Elementary Treatise on Artillery and Infantry, Adapted to the Service of the United States*. New York: G. P. Putnam, 1849.

127. Klass, Gert von. *Krupps: The Story of an Industrial Empire*. London: Sidgwick and Jackson, 1954.

128. Kosciusko, Thaddeus. *Exercises for Garrison and Field Ordnance*. New York: Campbell and Mitchell, 1808.

129. ———. *Maneuvers of Horse Artillery*. New York, 1800.

130. Lallemand, Henri D. *Treatise on Artillery*. Translated by James Renwick. New York: James Renwick, 1820.

131. Landmann, Isaac. *Principles of Artillery*. London, 1808.

132. Larter, Harry C. "Materiel of the First American Light Artillery, 1808–1809." *Military Collector and Historian*. September 1952, 53–63.

133. ———. "On Horses of the U.S. Artillery, Early 1800's." *Military Collector and Historian*, September 1954, 65–67.

134. Lauerma, Matti. *L'Artillerie de Campagne Française pendant les Guerres de la Révolution*. Helsinki: Soumalainen Tiedeakatemiaf, 1958.
135. LeBourg, J. H. *Essai sur l'Organisation de l'Artillerie et son Emploi la Guerre de Campagne*. Paris: J. Correaud, 1836.
136. Longacre, Edward G. *The Man behind the Gun: A Biography of General Henry Jackson Hunt, Chief of Artillery, Army of the Potomac*. New York: A. S. Barnes and Company, 1977.
137. Lyndall, Thomas. *Rifled Ordnance: A Practical Treatise on the Application of the Principle of Rifle to Guns and Mortars of Every Caliber*. New York, 1864.
138. Lynn, John A. *The Bayonets of the Republic: Motivation and Tactics of the Army of Revolutionary France, 1791–1794*. Urbana, IL: University of Illinois Press, 1984.
139. Mahon, John K. *The War of 1812*. Gainesville, FL: University of Florida Press, 1972.
140. Mallet, Robert. *The Construction of Artillery*. London: Longman Green, 1856.
141. Manchester, William. *The Arms of Krupp*. Boston: Little, Brown, 1964.
142. Manucy, Albert. *Artillery through the Ages: A Short Illustrated History of Cannon, Emphasizing Types Used in America*. Washington, DC: Government Printing Office, 1949.
143. May, Sir Edward S. *Achievements of Field Artillery*. Woolwich, England: Royal Artillery Institution, 1893.
144. Mazé, L. F. *État Actuel de l'Artillerie de Campagne en Europe*. 1838–1849.
145. ———. *Artillerie de Campagne en France, Description de l'Organisation et du Matériel de cette Arme en 1845*. Paris, 1845.
146. McElwee, William *The Art of Warfare: Waterloo to Mons*. Bloomington, IN: Indiana University Press, 1974.
147. McNeill, William H. *The Pursuit of Power: Technology, Armed Forces, and Society since* A.D. *1000*. Chicago: University of Chicago Press, 1982.
148. McPherson, James M. *Battle Cry of Freedom: The Civil War Era*. New York: Oxford University Press, 1988.
149. McWhiney, Grady, and Perry D. Jamieson. *Attack and Die: Civil War Military Tactics and the Southern Heritage*. University, AL: University of Alabama Press, 1982.
150. Menne, Bernard. *Krupp; or, The Lords of Essen*. London: Hodge and Company, 1937.
151. Mention, Léon. *L'Armée de l'ancien Régime de Louis XIV à la Révolution*. Paris: L. Henry May, 1900.
152. Mercer, Philip. *The Life of Gallant Pelham*. Macon, GA: J. W. Burke, 1929.
153. Monroe, John A. *The Rhode Island Artillery at the First Battle of Bull Run*. Providence, RI: 1878.
154. Moore, Edward A. *The Story of a Cannoneer under Stonewall Jackson*. Lynchburg, VA: J. P. Bell and Company, 1910.
155. Mordecai, Alfred. *Artillery for the Land Service of the United States*. Washington, DC: Government Printing Office, 1849.
156. ———. *Military Commission to Europe in 1855 and 1856*. Washington, DC: G. W. Bowman, 1860.
157. ———. *Reports on Experiments with Gunpowder at Washington Arsenal in 1843 and 1844*. Washington, DC: J. and G. S. Gideon, 1845.

158. ———. *Second Report on Experiments on Gunpowder Made at Washington Arsenal in 1845, 1847, 1848*. Washington, DC: J. and G. S. Gideon, 1849.
159. ———. *The Ordnance Manual for the Use of Officers of the US Army*. Washington, DC: Gideon and Company, 1850.
160. Morton, John W. *Forrest's Artillery*. Nashville, 1909.
161. Muller, John. *A Treatise of Artillery*. 2nd ed. London: J. Millan, 1768.
162. Naisawald, L. Van Loan. *Grape and Canister: The Story of the Field Artillery of the Army of the Potomac, 1861–1865*. New York: Oxford University Press, 1960.
163. Norton, Charles B. *Report on the Munitions of War*. Washington, DC: Government Printing Office, 1868.
164. Neese, George M. *Three Years in the Confederate Horse Artillery*. New York: Neal Publishers, 1911.
165. Owens, Charles H. and T. L. Dames. *Elementary Lectures on Artillery as Delivered at the Royal Military Academy*. Woolwich, England: John Boddy, 1859.
166. Owens, Charles H. *Principles and Practice of Modern Artillery, Including Artillery Materiel, Gunnery, and Organization and Use of Artillery in Warfare*. London: John Murray, 1871.
167. Patten, George. *Artillery Drill*. New York, 1863.
168. Paulding, J. N. *The Cannon and Projectiles Invented by Robert Parker Parrott*. New York: J. N. Paulding, 1879.
169. Peterson, Harold. *Round Shot and Rammers*. Harrisburg, PA: Stackpole Books, 1969.
170. Pottinger, J. *A Plain Treatise on Artillery*. Ahmednuggur, India, 1859.
171. Poumet, M. *Analyse et Dissertation Relative à la Nouvelle Artillerie*. Paris: Fain, 1827.
172. Poumet, P. A. *Instruction sur l'Artillerie de Campagne*. Paris, 1824.
173. Quimby, Robert S. *The Background of Napoleonic Warfare: The Theory of Military Tactics in Eighteenth Century France*. New York: AMS Press, 1957.
174. Ripley, Warren. *Artillery and Ammunition of the Civil War*. New York: Van Nostrand, 1970.
175. Roberts, Joseph. *The Hand-book of Artillery for the Services of the United States*. New York, 1863.
176. ———. *The Handbook of Artillery*. Richmond: J. W. Randolph, 1861.
177. ———. *The Handbook of Artillery for the Service of the U.S. Army and Militia*. New York: D. Van Nostrand, 1860.
178. Rodman, Thomas J. *Reports of Experiments on the Properties of Metals of Cannons and the Qualities of Cannon Powder with an Account of the Fabrication and Trials of a 15-inch Gun*. Boston: Charles H. Crosby, 1861.
179. Rothenberg, Gunther E. *The Art of Warfare in the Age of Napoleon*. Bloomington, IN: Indiana University Press, 1978.
180. Rouvroy, F. G. *Vorlesungen über die Artillerie*. 3 vols. Dresden, 1811–1814.
181. Scharnhorst, Gerhard von. *Handbuch der Artillerie*. 3 vols. Hannover, 1804–1814.
182. Schenkl, J. P. *Description of the Combination and Concussion Fuze*. Boston: John Wilson and Son, 1862.
183. Scoffern, John. *New Resources of Warfare with Especial Reference to Rifled Ordnance in their Chief Known Varieties*. London: Longman, Brown and Company, 1859.

183A. Showalter, Dennis E. *Railroads and Rifles: Soldiers Technology, and the Unification of Germany*. Hamden, CT: Archon Books, 1975.

184. Smith, Amasa. *A Short Compendium of the Duty of Artillerists*. Worcester, MA: 1800.

185. Smith, Justin H. *The War with Mexico*. 2 vols. New York: Macmillan, 1919.

186. Smith, Merritt Roe. *Harpers Ferry Armory and the New Technology: The Challenge of Change*. Ithaca, NY: Cornell University Press, 1977.

187. Stark, A. W. *Instruction for Field Artillery*. Richmond, VA: A. Morris, 1864.

188. Stevens, William. *A System for the Discipline of the Artillery of the United States of America*. New York: William H. Davis, 1797.

189. Strachan, Hew. *European Armies and the Conduct of War*. Boston: George Allen and Unwin, 1983.

190. Susane, General. *Histoire de l'Artillerie Française*. Paris: J. Hatzel, 1874.

191. Taylor, Franck. *Rifled Field Pieces: A Short Compilation of What Is Known of the New Field Artillery in Europe with Some Accounts of Our Own*. Washington, DC: 1862.

192. Tennant, J. Emerson. *The Story of the Guns*. London: Longman, Green, Longman, Roberts, and Green, 1864.

193. Thiroux, Laguionie. *Instruction Théorique et Pratique de'Artillerie*. Paris: Gaultier, 1837.

194. Thomas, Dean S. *Cannons: An Introduction to Civil War Artillery*. Gettysburg, PA: Thomas Publications, 1985.

195. Thomas, Lyndall. *Rifled Ordnance: A Practical Treatise on the Application of the Principle of Rifle to Guns and Mortars of Every Caliber*. New York: D. Van Nostrand, 1854.

196. Tidball, John C. "Remarks upon the Organization, Command and Employment of Field Artillery during War; Based on Experiences of the Civil War, 1861–65." Unpublished manuscript, Morris Swett Technical Library, U.S. Army Field Artillery School, Fort Sill, OK.

197. Tousard, Louis de. *American Artillerist's Companion*. Philadelphia: C. and A. Conrad, 1809. Reprint ed., Westport, CT: Greenwood Press, 1969.

198. Treadwell, Daniel. "On the Construction of Hooped Cannon, Being a Sequel to a Memoir on the Practicability of Constructing Cannon of Great Caliber." *Memoirs of the American Academy of Arts and Sciences*, 1864.

199. U.S. Ordnance Department. *The Ordnance Manual for the Use of Officers of the United States Army*. Washington, DC: Government Printing Office, 1850.

200. ———. *The Ordnance Manual for the Use of Officers of the United States Army*. Philadelphia: Washington, Gideon, and Company, 1850.

201. Walker, Francis. "General Hancock and the Artillery at Gettysburg." In Robert Underwood and Clarence C. Buel's eds., *Battles and Leaders of the Civil War*. Vol. III, New York: The Century Co., 1887–1888.

202. Wallace, Willard M. *Appeal to Arms: A Military History of the American Revolution*. Chicago: Quadrangle Books, 1951.

203. War Department (U.S.). *A System of Exercise and Instruction of Field Artillery Including Maneuvers for Light or Horse Artillery*. Boston: Hilliard, Gray, Little, Watkins, 1829.

204. ———. *Evolutions of Field Batteries of Artillery*. Translated by Robert Anderson. New York: D. Van Nostrand, 1860.

205. ———. *Instruction for Field Artillery*. Philadelphia: J. B. Lippincott and Company, 1860, 1861, 1863, 1864.
206. ———. *Instruction for Field Artillery*. Washington, DC: Government Printing Office, 1864.
207. ———. *Instruction for Field Artillery: Horse and Foot*. Baltimore: Joseph Robinson, 1845.
208. ———. *Instruction for Field Artillery: Horse and Foot*. Baltimore: Joseph Robinson, 1850.
209. Ward, Christopher. *The War of the Revolution*. Edited by John R. Alden. 2 vols. New York: Macmillan and Company, 1952.
210. Weller, Jac. "The Field Artillery of the Civil War." *Military Collector and Historian*, June 1953, 29–34.
211. Wiard, Norman. *Wiard's System of Field Artillery*. New York: Holman, 1863.
212. Wilkinson-Latham, Robert. *British Artillery on Land and Sea: 1790–1820*. Newton Abbott: David and Charles, 1973.
213. ———. *Napoleon's Artillery*. London: Osprey Publishing Ltd., 1975.
214. Wilson, A. W. *The Story of the Gun*. Woolwich, England: Royal Artillery Institution, 1944.
215. Winter, Frank H. *The First Golden Age of Rocketry: Congreve and Hale Rockets of the 19th Century*. Washington, DC: Smithsonian Institution, 1990.
216. Wise, Jennings C. *The Long Arm of Lee*. Lynchburg, VA: J. P. Bull Company, 1959.
217. Woodham-Smith, Cecil. *The Reason Why*. New York: E. P. Dutton and Company, 1953.

7

Introduction of Modern Field Artillery, 1866–1918

GENERAL HISTORIES

Besides the general studies identified in the first bibliographical essay, several key works on the wars between 1866 and 1918 also provide an invaluable background for understanding field artillery tactics, doctrine, organization, and materiel. For the Franco-Prussian War consult Michael Howard's *The Franco-Prussian War* (112). It is the standard work on the war and should be a point of departure for any serious study. John Friedrich's *The Franco-German War* (78) is also solid. Walter Millis's *The Martial Spirit* (146), H. Wayne Morgan's *America's Road to Empire: The War with Spain and Overseas Expansion* (152), and David F. Trask's *The War with Spain in 1898* (207) are important works on the Spanish-American War of 1898. The best modern account of the South African War is Thomas Pakenham's *The Boer War* (166). Another solid study on the Boer War is Byron Farwell's *The Great Anglo-Boer War* (72), as is L. S. Amery's (ed.) *The Times History of the War in South Africa* (5). The latter is replete with photographs and illustrations. Although it is old, G.F.R. Henderson's *The Science of War* (99) explains the impact of the South African War of 1899–1902 on the British army, as well as offering the writer's thoughts on tactics. For recent analysis of the British army during the latter years of the nineteenth century see Brian Bond's (ed.) *Victorian Military Campaigns* (30). Denis A. Warner's *The Tide at Sunrise: A History of the Russo-Japanese War* (231) is a solid account for that pivotal war.

There is a wealth of literature on World War I, but only a few key works will be cited here. Cyril B. Falls's *The Great War* (69) and James L. Stokesbury's *A Short History of World War I* (198) are starting places for the Great War. B. H. Liddell Hart's *The Real War: 1914–1918* (134) is a vital overview of the World War I.

FIELD ARTILLERY HISTORIES

Many outstanding books are available on field artillery developments during the years under consideration. William H. McNeill's *The Pursuit of Power: Technology, Armed Forces, and Society since* A.D. *1000* (139) places the introduction of rifled steel field artillery during the last decades of the nineteenth century in the context of a giant arms race among the European powers. Joseph Jobe's (ed.) *Guns: An Illustrated History of Artillery* (117) is a general survey of artillery history with an extensive coverage of field artillery developments during the last four decades of the nineteenth century and first two decades of the twentieth century. Another general survey of field artillery, Frank E. Comparato's *Age of Great Guns: Cannon Kings and Cannoneers Who Forged the Firepower of Artillery* (52) also has a solid analysis of rapid-fire field artillery tactics and technology at the turn of the twentieth century. O.F.G. Hogg's *Artillery: Its Origin, Heyday, and Decline* (108) is a general history of artillery with a useful section on rapid-fire field artillery. Even though it furnishes extensive coverage of field artillery tactics, doctrine, organization, and materiel in the twentieth century, Jonathan B. A. Bailey's *Field Artillery and Firepower* (9) is an invaluable work on field artillery at the turn of the century and has a useful bibliography. Bailey's book is, perhaps, the best available book on the evolution of field artillery since the introduction of rifled breechloaders. Kenneth Allen's *Big Guns of the Twentieth Century* (2) focuses on heavy artillery, but at the same time he furnishes insights into field artillery that are worth considering. Curt Johnson's *Artillery: The Big Guns Go to War* (118) is a short but solid study with illustrations and diagrams, while Ian V. Hogg's well-illustrated *The Guns: 1914–1918* (107) furnishes an excellent review of World War I field artillery.

John Batchelor and Ian V. Hogg's *Artillery* (12) covers various kinds of artillery and includes a section on rapid-fire field artillery. More anecdotal than the rest, Fairfax Downey's *Cannonade: Great Artillery Actions of History* (60) furnishes eloquent descriptions of field artillery operations. A brief but solid work on field artillery materiel is the U.S. Army Field Artillery School's *History of the Development of Field Artillery Materiel* (213). Reprinted several times, the book covers ammunition, fuzes, cannon design, and equipment.

During the latter years of the nineteenth century and early years of the twentieth century, contemporary histories of field artillery appeared. In *How Europe Armed for War* (157) J. T. Newbold analyzes European rearmament with rifled breechloading field artillery and small arms prior to World War I. Henry A. Bethell's *Modern Artillery in the Field* (20) is an excellent discussion of field artillery materiel, organization, and tactics at the beginning of the twentieth century. W. W. Greener's lengthy *The Gun and Its Development* (87) covers the transition from smoothbore to rifled artillery well. Heinrich Rohne's *Progress of Modern Field Artillery* (179), translated by Montgomery M. Macomb, is the best short history of the field artillery during the last three decades of the nineteenth century. Rohne covers tactics, technology, and organization. Another critical work, E. W. Lloyd and A. G. Hadcock's *Artillery: Its Progress and Present Position* (136) discuses the emergence of rapid-fire field artillery. Frederick G. Herr's *Artillerie* (Artillery, 100) is worth reading for the author's insights into field artillery history and the direction in which field artillery was heading following World War I.

In *Achievements of Field Artillery* (142) Sir Edward S. May explores the evolution of field artillery during the nineteenth century and the rise of rifled breechloaders during the latter years of the century, with an emphasis on the field artillery's role in combat. The same author's *Field Artillery with the Other Arms: Its Employment, Illustrated from Military History, and Its Re-armament with Quick-Firing Guns Discussed* (143) is worth examining for its analysis of combined arms warfare in the nineteenth century. James A. Longridge's *The Artillery of the Future and the New Powers* (137) explores the emergence of rifled breechloaders and tries to project where the new technology will take the field artillery. Although Oliver L. Spaulding's *Notes on Field Artillery for Officers of All Arms* (195) focuses primarily on technology of the early 1900s, it contains invaluable information on early rifled breechloaders that was considered important for all officers to know. Spaulding's *Weapons and Munitions of War: Artillery Weapons* (197) should also be read. Louis A.H. Langlois's *Les Artilleries de Campagne de l'Europe en 1874* (European Field Artillery in 1874, 130) covers European field artillery as it existed in the 1870s before the appearance of rapid-fire field guns.

Contemporary studies are available for the wars of the late nineteenth century. Sir John DuCane's *The Campaign in Alsace, August 1870* (63) is solid for field artillery operations during the initial days of the Franco-Prussian War. Ernst von Hoffbrauer's *The German Artillery in the Battles near Metz* (105) is critical for field artillery operations during the Franco-Prussian War from the German perspective. For another German viewpoint on the tactical lessons of the Franco-Prussian War read A. von Boguslawski's *Tactical Deductions from the War of 1870–1871* (29). Jean G.M. Rouquerol's *L'Artillerie dans le Bataille du 18 Aout* (The Artillery in the Battle of the 18th of August, 182) discusses French field artillery during the Franco-Prussian War of 1870–1871 from the French perspective.

For other wars involving European powers see Sir Edward S. May's *A Retrospect on the South African War* (141). It is excellent for its coverage of field artillery operations during the Boer War from the British perspective, as is D. F. Nicholls's *Our Artillery as Used in the Present War in South Africa and concerning Heavy Guns and Mountings in Particular* (158). F. V. Greene's *Report of the Russian Army and Its Campaigns in Turkey, 1877–1878* (86) should be consulted for Russian field artillery in the Russo-Turkish War. An old study on British artillery operations is *Military History of the Campaign of 1882 in Egypt* (206) by the War Office, Great Britain. Although it is old, Louis A.H. Langlois's

Lessons from Two Recent Wars (132) covers the field artillery lessons of the last wars of the nineteenth century. The French-language title is *Enseignements de Deux Guerres* (129).

Covering combined arms warfare during the last decades of the nineteenth century, Edward A. Altham's *The Principles of War Historically Illustrated* (3) has critical insights on field artillery operations and incorporates the lessons of the Russo-Japanese War into its examination of field artillery tactics. Ludwig von Estorff's *Taktische Lehren aus dem Russisch-Japanischen Feldkriege im Lichte unserer neuesten Vorschriften* (Tactical Lessons from the Russo-Japanese War as Illuminated in Our New Regulations, 68) should also be read for field artillery operations during the Russo-Japanese War and the lessons that the German army drew from them. Bruce I. Gudmundsson's *On Artillery* (90A) recounts the story of field artillery in the twentieth century and it's impact on the major conflicts of our time. Unique to this book are the discussions of issues that are usually papered over in the literature of field artillery—losses from "friendly fire," the frequent impotence of counter-battery fire, and the French origins of current American doctrine.

Of particular interest for World War I is C.N.F. Broad's "The Development of Artillery Tactics, 1914–1918" (35). For another invaluable perspective on field artillery in World War I see Alan F. Brooke's eight-part series entitled "The Evolution of Artillery in the Great War" (36). R. M. St. G. Kirke's "Some Aspects of Artillery Development during the First World War on the Western Front" (122) should be read, while R. M. Ogorkiewicz's "The Evolution of Self-propelled Guns" (164) is an essential overview of the early years of self-propelled artillery, which coincide with World War I. Sir Martin Farndale's *History of the Royal Regiment of Artillery, Western Front, 1914–1918* (70) covers British field artillery well during the Great War. Neil Fraser-Tytler's *Field Guns in France* (77) provides excellent insights about a British forward observer's abilities in World War I to shift fire around the battlefield.

Several books discuss the role of German artillery during World War I. They are Timothy T. Lupfer's *The Dynamics of Doctrine: The Changes in German Tactical Doctrine during the First World War* (138) and Bruce I. Gudmundsson's *Stormtroop Tactics: Innovation in the German Army, 1914–1918* (90). Georg Bruchmüller's *The German Artillery in the Break-Through Battles of the World War* (39) is a thoughtful examination of German field artillery performance. The U.S. Army, American Expeditionary Force's *Notes on German Artillery Materiel* (212) covers German field artillery in World War I well. David Nash's brief, illustrated *German Artillery: 1914–1918* (155) is a solid account of materiel and equipment.

General Gascouin's *L'Évolution de l'Artillerie pendant la Guerre* (The Artillery's Evolution during the War, 79) is a critical examination of French field artillery operations during World War I. Another helpful work on French field artillery in World War I is Frédéric Culmann's *Tactique d'Artillerie* (Artillery Tactics, 56). Frederick Herr's *The General Report of General Herr* (101) is worth reading for the General's insights. J. Campana's *Les Progrès de L'Artillerie* (The Artillery's Progress, 47) is an exceptional study of French field artillery materiel, tactics, organization, and operations during World War I.

William Crozier's *Ordnance and the World War: A Contribution to the History of American Preparedness* (55) explains the dilemma of the United States's effort to arm for World War I. Also see Benedict Crowell's *America's Munitions: 1917–1918* (54). Edward S. Farrow's *American Guns in the War with Germany* (71) is an excellent source. William J. Snow's (q.v.) *History of the Office, Chief of Field Artillery* (193) is critical for its examination of the office during World War I and its contribution to developing the U.S. Army's officer corps and enlisted ranks. Snow's *Signposts of Experience: World War Memoirs of Major General William J. Snow* (194) is also essential for understanding American field artillery developments during World War I as seen from the viewpoint of General Snow, who was the Chief of Field Artillery during the Great War and through most of the 1920s.

Two useful memoirs of World War I artillerymen exist. William Carr's *A Time to Leave the Ploughshares: A Gunner Remembers, 1917–1918* (49) is a useful memoir about a Scottish farmer, who was drafted in 1916, selected for officer training, and received his commission as an officer in the Royal Artillery. Herbert Sulzbach's *With the German Guns: Four Years on the Western Front, 1914–1918* (200), initially published in German in 1935 as *Zwei Lebende Mauern*, is based on the author's diaries that he kept throughout the war and furnishes a view of an artilleryman's experience of trench warfare.

For British field artillery consult Shelford Bidwell's *Gunners at War: A Tactical Study of the Royal Artillery in the Twentieth Century* (22) for its perceptive insights into tactics, doctrine, and operations. Kenneth Brookes's *Battle Thunder: The Story of British Artillery* (37) is particularly valuable for general coverage of British field artillery in the twentieth century. Shelford Bidwell and Dominick Graham's *Firepower: British Army Weapons and Theories of War, 1904–1945* (23) is the best account of British field artillery tactics, organization, and materiel during the first four decades of the twentieth century. It also has an outstanding bibliography. C.A.L. Graham's *History of the Indian Mountain Artillery* (84) and *The Story of the Royal Regiment of Artillery* (85) are valuable. Sir Charles E. Caldwell's two-volume *The History of the Royal Artillery: 1860–1914* (46) is worth examining for insights into the British army's adoption of rapid-fire field artillery and the adjustments that it caused. H.W.L. Hime's *History of the Royal Artillery* (103) is solid, while O.F.G. Hogg examines field artillery manufacturing in *The Royal Arsenal: Its Background, Origin, and Subsequent History* (109). Ubique's *Famous Battles of the Royal Artillery* (211) is a sound operational history.

For German field artillery consult Rolf Hinze's *Die Hannoverische Artillerie* (Hanoverian Artillery, 104). In his study Hinze devotes attention to German field artillery from 1848 to 1945. Friedrich Feeser's *Artillerie im Feldkrieg* (Artillery in War, 75) covers German field artillery history from 1600s to 1918.

There are also useful studies on French field artillery that should be consulted.

Charles A. Romain's *Les Responsabilités de l'Artillerie Française en 1870* (The Responsibilities of French Artillery in 1870, 180) provides an invaluable review of French field artillery as it converted from smoothbore to rifled weapons. Another outstanding work is François Rimailho's *Artillerie de Campagne* (Field Artillery, 174). He covers French field artillery from 1870 through 1914 with especially sound coverage of the introduction of the M1897 75mm field gun, which revolutionized field artillery.

Christopher Bellamy's "The Russian Artillery and the Origins of Indirect Fire" (15) is a key two-part article on the development of indirect fire during the latter years of the nineteenth century, especially as it relates to the Russian army. For a general history of Russian and Soviet field artillery see Christopher Bellamy's "The Firebird: 600 Years of the Russian Artillery" (13) and *Red God of War: Soviet Artillery and Rocket Forces* (14). The latter is, perhaps, the best work devoted solely to Soviet and Russian field artillery that is available in the English language.

Boyd L. Dastrup's *King of Battle: A Branch History of the U.S. Army's Field Artillery, 1775–1980* (58) is a solid starting point for the study of American field artillery between 1866 and 1918 as it converted from direct fire to indirect fire. Although it is short, Tom Jones and Fitzhugh McMaster's "Light Battery F, 5th Regiment of Artillery, Peking, 1900" (120) is interesting.

For studies on key personalities see G. Klass's *Krupps: The Story of an Industrial Empire* (123), Wilhelm Berdrow's *The Krupps: 150 Years of Krupp History* (17), and Bernhard Menne's *Krupp; or, The Lords of Essen* (144). William Manchester's *The Arms of Krupp, 1587–1968* (140) is controversial but worth reading because of its analysis of the rise of the Krupp family in the arms industry. J. D. Scott's *Vickers* (191) and Clive A. Trebilcock's *The Vickers Brothers: Armaments and Enterprise, 1854–1914* (208) are critical works. Also see David Dougan's *The Great Gunmaker: The Story of Lord Armstrong* (59). For French field artillery manufacturers read Joseph A. Roy's *Histoire de la Famille Schneider et du Creusot* (History of the Schneider Family and of Creusot, 185).

Although they may be more difficult to obtain, theses and dissertations are also worth examining. Robert S. Ballagh's "The Development of French Field Artillery Doctrine, 1870–1914: Technological Change and Tactics" (11) is noteworthy for its discussion of the French transition from direct to indirect fire. Robert H. Scales's "Artillery in Small Wars: The Evolution of British Artillery Doctrine, 1860–1914" (188) should be considered. For American field artillery see Larry D. Roberts's "The Artillery with the Regular Army in the West" (176), Wallace J. Savoy's "The Evolution of the American Modern Light Field Gun" (187), and Vardell E. Nesmith's "The Quiet Paradigm Change: The Evolution of the Field Artillery Doctrine of the United States Army, 1861–1905" (156). Savoy's and Nelsmith's works provide invaluable coverage of the transition from smoothbore to rifled field artillery.

Professional journals are also valuable sources of information for the years from 1866 through 1918. For the debates over tactics, materiel, and organization consult *Field Artillery Journal* (United States), *Journal of the Military Service Institution of the United States*, *Army and Navy Journal* (United States), *Journal of the United States Artillery*, *Military Review* (United States), *Cavalry/Armor Journal* (United States), *Journal of the Royal United Service Institution*, *Proceedings of the Royal Artillery Institution/Journal of Royal Artillery*, *Journal des Sciences Militaires* (France), *Revue d'Artillerie* (France), *Militär Wochenblatt* (Germany). *Field Artillery Journal*, *Journal of the Military Service Institution*, *Military Review*, *Cavalry/Armor Journal*, *Journal of the Royal United Service Institution*, and *Proceedings of the Royal Artillery Institution/Journal of Royal Artillery* are indexed to assist the serious researcher.

TECHNOLOGY

For a general consideration of European field artillery prior to the adoption of rifled steel breechloaders see Charles B. Brackenbury's *European Armaments in 1867* (32). Another work on early breechloaders is Charles B. Norton and W. J. Valentine's *Report to the Government of the United States on the Munitions of War Exhibited at the Paris Universal Exhibition, 1867* (163).

Sir Andrew Nobles's *Artillery and Explosives: Essays and Lectures Written and Delivered at Various Times* (162) provides invaluable insights into the technology at the turn of the twentieth century. Rudolf H. C. Verner's *Guns and Projectiles* (218) explores field artillery materiel during the first years of the twentieth century also. Ormond Lissak's *Ordnance and Gunnery* (135) discusses the state of field artillery technology at the turn of the century. Charles C. Morrison's *Modern Guns and Mortars* (153) also covers European field artillery materiel, as does John G. Butler's *Projectiles and Rifled Cannons: Systems of Projectiles and Rifling, with Practical Suggestions for Their Improvement as Embraced in a Report to the Chief of Ordnance, United States Army* (45). Henry A. Bethell's *Modern Guns and Gunnery* (21) is the best work on field artillery at the beginning of the twentieth century. He examines materiel and tactics. Another solid study of early rapid-fire field artillery materiel is Sisson C. Pratt's *Field Artillery: Its Equipment, Organization, and Tactics* (173). Charles H. Owens's *The Principles and Practice of Modern Gunnery, Including Artillery Materiel, Gunnery, and Organization and Use of Artillery in Warfare* (165) addresses artillery materiel, gunnery, organization, and tactics. The U.S. Army War College's *Field Artillery Notes* (215) is worth examining for its information on European and American field artillery materiel.

For World War I technology examine Sevellon Brown's *The Story of Ordnance in the World War* (38) and Aubrey Wade's *The War of the Guns: Western Front, 1917 and 1918* (219). The U.S. Ordnance Department's *Handbook of Ordnance Data* (217) is particularly helpful for European artillery technology during World War I.

Douglas T. Hamilton's *Shrapnel Shell Manufacture* (95) is a comprehensive treatise on the forging, machining, and heat treatment of shells and the manufacture of cartridge cases and fuses for shrapnel employed for field and mountain artillery. The same author's *High-Explosive Shell Manufacture* (94) is also vital, while John B. Bernadou examines high explosives in *Smokeless Powder, Nitrocellulose, and the Theory of the Cellulose Molecule* (18). Oscar Guttman's two-volume *The Manufacture of Explosives* (91) should also be considered, as should Marcellin Berthelot's *Explosives and Their Power* (19), translated by C. Napier Hake and William Macnab. For a general bibliography of high explosives consult H. Brunswig's *Explosives: A Synoptic and Critical Treatment of the Literature of the Subject as Gathered from Various Sources* (41), translated by Charles E. Munroe and Alton L. Kibler. The science of ballistics is covered in James M. Ingalls's *Interior Ballistics* (115) and *Ballistics* (114).

For French field artillery technology see Edmond A. L. Buat's *The Field Artillery: Its History, Evolution and Present Status* (42), translated by Felix Alcan. It is valuable for materiel, organization, and tactics at the beginning of the twentieth century. Another worthwhile book on French field artillery is Jules M. J. Challéat's *L'Artillerie de Terre en France: 1816–1919* (French Artillery, 1816–1919, 50), while James E. Hicks's *Notes on French Ordnance: 1717–1936* (102) has solid coverage of rapid-fire field artillery technology. A more sophisticated analysis of French field artillery is Sébastien Léon Wisse's *A la Gloire de l'Artillerie Française de 1870 à 1918* (The Glory of French Artillery from 1870 to 1918, 236), while Heinrich Rohne's *L'Artillerie de Campagne Française* (French Field Artillery, 178) is another dependable work on French field artillery at the turn of the century. Ignace Campana's *L'Artillerie de Campagne, 1792–1901* (Field Artillery, 1792–1901, 48) is an account of the introduction of rapid-fire field artillery in the French army as well as earlier field artillery technology, while E. Monthaye's *Krupp and de Bange* (148), translated by O. E. Michaelis, compares the great breech-loading cannons of Europe. James Dredge's (ed.) *Modern French Artillery: The St. Chamond, De Bange, Canet, and Hotchkiss Systems* (61) is an invaluable comparative study written by British ordnance experts on field artillery systems in France during the latter decades of the nineteenth century. For a short analysis of the French M1897 75mm field gun see M. Th. Schloesing's *The French "75"* (190), translated by Frank H. Canaday. For outstanding information on the French M1897 75mm field gun consult Lucien J. R. André's *Notes on the French 75mm Gun Matériel* (6).

Étienne Girardon's *Leçons d'Artillerie, Conformes au Programme de l'École Militaire de l'Artillerie* (Artillery Lessons as Consistent with the Artillery School's Curriculum, 82) is especially sound on French rapid-fire artillery. He covers ordnance, tactics, and ballistics well. The same author examines French field artillery materiel at the turn of the century in *Organisation du Matériel d'Artillerie* (Artillery Organization and Materiel, 83).

Kraft Karlzu Hohenlohe-Ingelfingen's *Letters on Artillery* (110), translated by N. L. Walford, is a must for Prussian field artillery reforms during the years after the Franco-Prussian War of 1870–1871. R. Bille's *Das deutsche Feld-*

Artillerie Materiel vom Jahre 1873 (German Field Artillery Materiel from 1873, 24) covers only a few years, but it is valuable. First Lieutenant von Bleyhoffer's *Die schwere Artillerie des Feldheeres* (The Army's Heavy Artillery, 28) discusses German field artillery of the late nineteenth century and early twentieth century. The Prussian General Staff's *The Evolution of Modern Field Artillery* (81) is also excellent on materiel as is Captain von Roskaten's two-volume *Die heutige Feldartillerie mit Rohrücklauf: Ihr Materiel, technische Hilfsmittel, Schiessverfahren, Organisation, und Taktik* (Today's Field Artillery: Its Materiel, Equipment, Techniques of Shooting, Organization, and Tactics, 181). Roskaten also has a short section on field artillery history. Ernst von Hoffbauer's *Schwebende Feldartillerie Fragen: Taktischartilleristische Studie für Offiziere aller Waffen* (Field Artillery Questions: A Study of Artillery Tactics for Officers of all the Arms, 106) is a sound discussion of German artillery. Heinrich Rohne's *An Essay on the Shrapnel Fire of Field Artillery* (177), translated by N. L. Walford, is a theoretical examination of shrapnel fire from field guns. *Feld Artillerie der Zukunft* (Field Artillery of the Future, 73) tries to project the future of European and German field artillery organization, training, and weapons.

For British artillery examine Arthur Forbes's three-volume *A History of Army Ordnance Service* (76). It covers field artillery technology in the British army as well as other related topics. *Treatise on the Manufacture of Guns and Text Book of Ordnance in the British Service* (192) by the Secretary of State for War, Great Britain, is invaluable. Other worthwhile books on field artillery materiel are Frederick C. Morgan's *Handbook of Artillery Materiel* (151), Francis Stoney's *A Text Book of the Construction and Manufacture of the Rifled Ordnance in the British Service* (199), and J. J. Crooks's *History of the Royal Irish Artillery* (53).

Carlo Montu's thirteen-volume *Stòria dello Artiglierìa Italiana* (The History of Italian Artillery, 149) has a volume that covers 1870–1914. G.W.L. Nicholson's *The Gunners of Canada: The History of the Royal Regiment of Canadian Artillery* (159) is invaluable for Canadian field artillery.

For American field artillery read Harry G. Bishop's (q.v.) *Field Artillery: King of Battle* (27). This book provides invaluable insights into the U.S. Army's field artillery technology, tactics, and organization at the beginning of the twentieth century. Also see Alexander B. Dyer's *Handbook for Light Artillery* (64) for its solid coverage of field artillery materiel. *Organization of the Ordnance Department and the Reasons for Its Separation from the Line of the Army* (16), by Stephen Vincent Benét, is a brief history of the reasons behind separating the Ordnance Department from the line. For a solid history of field artillery technological developments during the last decades of the nineteenth century in the United States see *Gun Making in the United States* (25), written by Rogers Birnie, Jr. For a general examination of American ordnance see Henry Metcalfe's *A Course of Instruction in Ordnance and Gunnery* (145). The United States War Department's *Modern Guns and Gunnery: A Practical Manual for Officers of*

Horse, Field, and Mountain Artillery (229) outlines American field artillery during the first decade of the twentieth century. Because of its importance, this book went through three editions. Another sound book is Harry G. Bishop's *Elements of Modern Field Artillery* (26). He stresses American and European ammunition, equipment, cannons, and tactics. William I. Westervelt's *Gunnery and Explosives for Field Artillery Officers* (233) is a reference book on American field artillery.

Designed as a textbook for the U.S. Military Academy, William H. Tschappat's *Textbook of Ordnance and Gunnery* (210) is an excellent source for American ordnance and gunnery prior to World War I. The U.S. Ordnance Department's *Handbook of Ordnance Data* (217) is invaluable. T.T.S. Laidley's *The Ordnance Department of the U.S. Army* (128) is short but worth the time reading. Intended for cadets at the U.S. Military Academy, Lawrence L. Bruff's *A Text-Book of Ordnance and Gunnery: Prepared for the Use of Cadets of the U.S. Military Academy* (40) is an overview of gunnery and ordnance in the U.S. Army at the beginning of the twentieth century. The U.S. Ordnance Department's *Collection of Annual Reports and Other Important Papers Relating to the Ordnance Department, 1812–1889* (216) provides key documents on the development of field artillery materiel in the United States. The U.S. Army Field Artillery School's *Materiel* (214) discusses the evolution of American and European field artillery and is particularly valuable for its discussion of the development of rifled breechloaders. L. P. Alford's *Manufacture of Artillery Ammunition* (1) provides valuable insights into the production of ammunition. One of the best sources on the development of American rapid-fire field artillery is Charles B. Wheeler's (q.v.) *A Discussion of the Proposed Methods to Increase the Rapidity of Fire of Field Guns* (234).

TACTICS, DOCTRINE, AND ORGANIZATION

Paddy Griffith's *Forward into Battle: Fighting Tactics from Waterloo to Vietnam* (89) is an excellent study of tactics and a starting point. More narrow in scope, L. F. Haber's *The Poisonous Cloud: Chemical Warfare in the First World War* (92) covers gas warfare in World War I well.

Contemporary writings on tactics are abundant for research. For French field artillery tactics see Charles Kessler's *Tactique des Trois Armes* (Tactics of the Three Arms, 121). Hippolyte I. Plessix's *Manuel Complète d'Artillerie* (The Complete Manual of Artillery, 172) covers field artillery tactics. One of the best works is Gabriel Rouquerol's *The Tactical Employment of Quick-firing Field Artillery* (183), translated by P. De B. Radcliffe. Another key French work is Ernest Jaegle's *L'Artillerie* (The Artillery, 116) because of its examination of the impact of the new field artillery technology upon tactics and organization during the 1880s. Gaston P. A. Aubrat's *Les Exercices de Service en Campagne dans le Gruppe de Batteries* (Drilling Groups of Batteries for Combat, 7) is an essential work on French field artillery tactics at the beginning of indirect fire.

Hippolyte Langlois's three-volume *L'Artillerie de Campagne en Liaison avec les Autres Armes* (Field Artillery with the Other Arms, 131) is invaluable for the author's insights into tactics. Henri A. Niessel's *Combinaison des Efforts de l'Infanterie et de l'Artillerie dans le Combat* (The Combined Action of Infantry and Artillery in Combat, 161) is another critical work on field artillery and infantry tactics. The Ministère de la Guerre's *Cours Special à l'Usage des Sous-Officers d'Artillerie* (Special Course for the Use of Junior Artillery Officers, 147). Frédéric Culmann's *Le Canon à Tir Rapide dans la Bataille* (The Quick-firing Gun in Battle, 57) is an excellent source on tactics. Alexis H. Brailmont's two-volume work *Tactique de Combat des Trois Armes* (Tactics of the Three Arms, 34) should also be considered. Alexandre Percin's *Cinq Années de Inspection* (Five Musters for the Year, 171) and *Artillery at the Picardy Maneuvers in 1910* (170) are worth reading for their critical and insightful comments on field artillery tactics.

Other enlightening sources on French field artillery tactics are Pierre Alvin's *Leçons d'Artillerie* (Artillery Lessons, 4), Henri L. LeBlond's *Le Canon à Tir Rapide et l'Instruction de l'Artillerie* (Rapid-fire Artillery and Instruction for the Artillery, 133), Jacques E.P. Feline's *L'Artillerie au Maroc* (The Artillery in Morocco, 74), and François A. Trequier's *L'Artillerie pour les Autres Armes* (Artillery Support for the Other Arms, 209). Jacques Rouvier's *Present-Day Warfare: How an Army Trains and Fights* (184) should also be considered when examining tactics. He covers combined arms warfare and has a useful chapter on field artillery tactics and weapons. Pierre Joseph Louis Alfred DuBois's *L'Artillerie de Campagne dans la Guerre Actuelle* (Field Artillery in Present Warfare, 62) is a treatise on field artillery tactics. Fenton J. Aylmer's *French Views on the Tactical Employment of Field Artillery* (8) should be considered. Henri J. Paloque's *Artillerie de Campagne* (Field Artillery, 167) and Samuel Bourquet's *L'Artillerie dans le Combat* (Artillery in Combat, 31) should also be considered when studying French field artillery tactics.

For German tactics before the introduction of rapid-fire field guns see the Kriegsministerium's *Drill Regulations of the German Field Artillery* (124), translated by W. A. Macbean. The Kriegsministerium's *Drill Regulations of the German Artillery* (125), translated by J.H.V. Crowe, should also be read. Another source of German field artillery drill and tactics before the introduction of rapid-fire breechloaders is the Kriegsministerium's *Entwurf zu einem Exerzier-Reglement für die Feldartillerie der königlichen Preussischen Armee* (Drill Regulations for Prussian Field Artillery, 126). A later work on pre-rapid-fire field artillery is Adolf von Schell's *The Tactics of Field Artillery* (189), translated by A. E. Turner. Schell's work is an excellent discussion of German field artillery tactics at the turn of the twentieth century. The Kriegsministerium's *Gefechtsvorschrift für die Artillerie* (Combat Instructions for the Artillery, 127) should be consulted for German field artillery tactics in World War I. One of the most outstanding treatises on German field artillery tactics is W. Balck's *Tactics*, Volume III (10). Major de Pardieu's *A Critical Study of German Tactics and of the New German Regulations* (168), translated by Charles F. Martin, discusses the introduction of indirect fire. Ritter von Eberhard's *Das Wesen der modernen Visiervorrichtungen der Landartillerie* (The Characteristics of Modern Sighting Equipment of Artillery, 65) is also worth reading for the German application of indirect fire.

For other sources on German field artillery tactics see the German General Staff's *Drill Regulations of the German Field Artillery* (80), published by His Majesty's Stationary Office in Great Britain, and J. M. Grierson's *Confidential Notes on the Training of German Field Artillery* (88). Additional works include von Wernigk's *Taschenbuch für die Feldartillerie* (Pocketbook for Field Artillery, 232) and R. Wille's *Granatschrapnell-Brisanzschrapnell* (Shrapnel and High-Explosive Shell, 235).

Charles E.D. Budworth's *Tactical Employment of Artillery as Evolved on the Practice Ground and from the Experience of Modern War* (44) furnishes a historical perspective on the development of field artillery tactics in Great Britain. For more on British tactics see Edward B. Hamley's *The Operations of War Explained and Illustrated* (97) and Henry M. Johnstone's *A History of Tactics* (119). Other helpful books are E. F. Hall's *Cooperation between Infantry and Artillery and the Best Means of Obtaining It* (93) and Alexander Hamilton-Gordon's *Changes in Field Artillery Equipment and Their Effect on the Action of Field Artillery in the Attack* (96). H. T. Russell's *The Employment of Artillery in the Field in Co-operation with the Other Arms* (186) is enlightening for its discussion of combined arms warfare as influenced by indirect fire, as is C. Francis Clery's *Minor Tactics* (51). Robert Homes's *A Précis of Modern Tactics* (111) and F. B. Elmslie's *The Possible Effects on Tactics of Recent Improvements in Weapons* (67) are also worthwhile. Charles E.D. Budsworth's *Artillery in Cooperation with Infantry* (43) explores field artillery and infantry operations. In *The Organization and Tactical Use of Cavalry, Artillery, and Infantry* (98) C. L. Harvey examines field artillery tactics during the last days of direct fire.

Additional works on British field artillery exist. See the War Office, Great Britain, *Field Artillery Service Handbook* (202), *Manual of Artillery Exercises* (205), *Field Artillery Drill* (201), *Field Artillery Training* (203), and *Instructions for Practice, Field Artillery* (204). All of them offer invaluable insights into British tactics, drill, and organization. Other books include Henry Brackenbury's *The Use and Abuse of Field Artillery* (33). Also consult A. L. Pemberton's *The Development of Artillery Tactics and Equipment* (169) for its useful information. For a look at British field artillery as rifled breechloaders were being introduced consult Garnet Wolseley's *The Soldier's Pocketbook* (237). For a discussion on battery operations see W. A. Nicholson's *Artillery Fire: The Battery* (160). W.B.E. Ellis's *Brief Notes on Field Artillery* (66) is a solid study. Arthur L. Hunt's *Artillery Lines of Fire* (113) is another worthwhile work on tactics.

Manuals for field artillery tactics in the U.S. Army are also available and plentiful. Joseph Roberts's *The Handbook of Artillery for the Service of the United States Army and Militia* (175) is a reprint of an earlier edition, but it is still worth examining. Other manuals on drill for American field artillery are the U.S. War Department's *Instruction for Field Artillery* (227), written by William French, William F. Barry (q.v.), and Henry J. Hunt (q.v.), and *Light Artillery Drill Regulations* (228). For another examination of drill and tactics prior to the introduction of quick-fire field artillery see the U.S. War Department's *Artillery Tactics* (221). Both editions of the manual discuss field artillery operations with cavalry and infantry. The U.S. War Department's *Drill Regulations for Light Artillery* (226) is the last pre–indirect fire drill and tactical manual for the American army. Charles C. Morrison and James C. Ayres's *Course of Instruction for Artillery Gunners: Modern Guns and Mortars Adopted in the United States Land Service, Their Carriages, Projectiles, Fuzes, and Sights* (154) is worth examining for gunnery and artillery equipment.

The U.S. War Department's *Drill Regulations for Field Artillery, U.S. Army (Provisional)* (222) introduces indirect fire into the U.S. Army's field artillery. The U.S. War Department's *Drill Regulations for Field Artillery, U.S. Army (Provisional)* (223) expands upon the 1905 drill manual, as does its *Drill Regulations for Field Artillery, U.S. Army* (224). For more information on American field artillery drill consult the War Department's *Drill Regulations for Field Artillery (Horse and Light)* (225) and *Provisional Drill and Service Regulations for Field Artillery (Horse and Light)* (230).

The most valuable work on American field artillery tactics at the turn of the century is Oliver L. Spaulding's *Notes on Field Artillery* (196). He covers materiel, gunnery, drill, and tactics. Because of its popularity, the book went through several editions during the early years of the twentieth century. For more on American battery organization and drill see Onorio Moretti's *Notes on Training Field Artillery Drill* (150). Arthur L. Wagner's *Organization and Tactics* (220) is a solid examination of American tactics, including field artillery at the turn of the century.

BIBLIOGRAPHY

1. Alford, L. P. *Manufacture of Artillery Ammunition*. New York: McGraw Hill Book Company, 1917.
2. Allen, Kenneth. *The Big Guns of the Twentieth Century*. Hove, East Anglia, England: Firefly Books, 1976.
3. Altham, Edward A. *The Principles of War Historically Illustrated*. London: Macmillan, 1914.
4. Alvin, Pierre. *Leçons d'Artillerie*. 2 vols. Paris: R. Chapelot, 1908.
5. Amery, L. S., ed. *The Times History of the War in South Africa*. London: S. Low Maiston and Company Ltd., 1900–1909.
6. André, Lucien J.R. *Notes on the French 75mm Gun Materiel*. Translated by U.S. War Department. Washington, DC: Government Printing Office, 1917.
7. Aubrat, Gaston P.A. *Les Exercices de Service en Campagne dans la Gruppe de Batteries*. Paris: Berger-Levrault, 1901.

8. Aylmer, Fenton J. *French Views on the Tactical Employment of Field Artillery*. London: Hugh Rees, 1911.
9. Bailey, Jonathan B.A. *Field Artillery and Firepower*. Oxford, England: The Military Press, 1989.
10. Balck, W. *Tactics*. Volume III. London: Hugh Rees, 1914.
11. Ballagh, Robert S. "The Development of French Field Artillery Doctrine, 1870–1914: Technological Change and Tactics." Doctoral dissertation, Duke University, 1971.
12. Batchelor, John, and Ian V. Hogg. *Artillery*. New York: Charles Scribner's Sons, 1972.
13. Bellamy, Christopher. "The Firebird: 600 Years of Russian Artillery." *History Today*, September 1982, 16–20.
14. ———. *Red God of War: Soviet Artillery and Rocket Forces*. Washington, DC: Brassey's Defence Publishers, 1986.
15. ———. "The Russian Artillery and the Origins of Indirect Fire." *Army Quarterly and Defense Journal*. Part I, April 1982, 211–22, and Part II, July 1982, 330–37.
16. Benét, Stephen Vincent. *Organization of the Ordnance Department and the Reasons for Its Separation from the Line of the Army*. Washington, DC: Government Printing Office, 1874.
17. Berdrow, Wilhelm. *The Krupps: 150 Years of Krupp History*. Berlin, 1937.
18. Bernadou, John B. *Smokeless Powder, Nitrocellulose, and the Theory of the Cellulose Molecule*. New York: John Wiley and Sons, 1901.
19. Berthelot, Marcellin. *Explosives and Their Power*. Translated by C. Napier Hake and William Macnab. London: John Murray, 1892.
20. Bethell, Henry A. *Modern Artillery in the Field*. London: Macmillan, 1911.
21. ———. *Modern Guns and Gunnery*. Woolwich, England: F. J. Cattermole, 1910.
22. Bidwell, Shelford. *Gunners at War: A Tactical Study of the Royal Artillery in the Twentieth Century*. London: Arms and Armour Press, 1970.
23. Bidwell, Shelford, and Dominick Graham. *Firepower: British Army Weapons and Theories of War, 1904–1945*. Boston: George Allen and Unwin, 1982.
24. Bille, R. *Das deutsche Feld-Artillerie Materiel vom Jahre 1873*. Berlin: Verlag von A. Bath, 1876.
25. Birnie, Rogers, Jr. *Gun Making in the United States*. Washington, DC: Government Printing Office, 1914.
26. Bishop, Harry G. *Elements of Modern Field Artillery*. Menasha, WI: George Banta Publishing Company, 1914.
27. ———. *Field Artillery: King of Battle*. Boston: Houghton Mifflin, 1935.
28. Bleyhoffer, First Lieutenant von. *Die schwere Artillerie des Feldheeres*. Berlin: R. Eisenschmidt, 1905.
29. Boguslawski, A. von. *Tactical Deductions from the War of 1870–1871*. Fort Leavenworth, KS: George A. Spooner, 1891.
30. Bond, Brian, ed. *Victorian Military Campaigns*. New York: Praeger, 1967.
31. Bourquet, Samuel. *L'Artillerie dans le Combat*. Paris: R. Chapelot, 1908.
32. Brackenbury, Charles B. *European Armaments in 1867*. London: Chapman and Hall, 1867.
33. Brackenburg, Henry. *The Use and Abuse of Field Artillery*. Aldershot, England: Gale and Polden, 1888.

34. Brailmont, Alexis H. *Tactique de Combat des Trois Armes*. Paris: L. Baudoin, 1881.
35. Broad, C.N.F. "The Development of Artillery Tactics, 1914–1918." *Journal of the Royal Artillery*, May 1922, 62–81, and June 1922, 127–28.
36. Brooke, Alan F. "The Evolution of Artillery in the Great War." *Journal of the Royal Artillery* 5 (1924):250–67; January 1925, 359–72; April 1925, 37–51; October 1925, 369–87; April 1926, 76–93; July 1926, 232–49; October 1926, 320–26; January 1927, 469–82.
37. Brookes, Kenneth. *Battle Thunder: The Story of British Artillery*. Reading, England: Osprey Publishing Ltd., 1979.
38. Brown, Sevellon. *The Story of Ordnance in the World War*. Washington, DC: James Williams, 1920.
39. Bruchmüller, Georg. *The German Artillery in the Break-Through Battles of the World War*. Translated by J. H. Wallace. Fort Sill, OK: Field Artillery School, 1922.
40. Bruff, Lawrence L. *A Text-Book of Ordnance and Gunnery: Prepared for the Use of Cadets of the U.S. Military Academy*. New York: John Wiley and Sons, 1896.
41. Brunswig, H. *Explosives: A Synoptic and Critical Treatment of the Literature of the Subject as Gathered from Various Sources*. Translated by Charles E. Munroe and Alton L. Kibler. New York: John Wiley and Sons, 1912.
42. Buat, Edmond A. L. *The Field Artillery: Its History, Evolution, and Present Status*. Translated by Felix Alcan. Paris: Felix Alcan, 1911.
43. Budsworth, Charles E.D. *Artillery in Cooperation with Infantry*. London: Hugh Rees, 1910.
44. ———. *Tactical Employment of Artillery as Evolved on the Practice Ground and from the Experience of Modern War*. London: Hugh Rees, 1908.
45. Butler, John G. *Projectiles and Rifled Cannons: Systems of Projectiles and Rifling, with Practical Suggestions for Their Improvement as Embraced in a Report to the Chief of Ordnance, United States Army*. New York: D. Van Nostrand, 1875.
46. Caldwell, Sir Charles E. *The History of the Royal Artillery: 1860–1914*. 2 vols. Woolwich, England: Royal Artillery Institution, 1931.
47. Campana, J. *Les Progrès de l'Artillerie*. Paris: Charles-Levrault, 1923.
48. ———. *L'Artillerie de Campagne, 1792–1901*. Paris: Henri Charles Lavauzelle, 1901.
49. Carr, William. *Time to Leave the Ploughshares: A German Remembers, 1917–1918*. London: Robert Hale, 1985.
50. Challéat, Jules M. J. *L'Artillerie de Terre en France: 1816–1919*. Paris: Imprimerie Nationale, 1933.
51. Clery, C. Francis. *Minor Tactics*. London: K. Paul, Trench and Company, 1886.
52. Comparato, Frank E. *Age of Great Guns: Cannon Kings and Cannoneers Who Forged the Firepower of Artillery*. Harrisburg, PA: The Stackpole Company, 1965.
53. Crooks, J. J. *History of the Royal Irish Artillery*. Dublin: Browne and Nolan, 1914.
54. Crowell, Benedict. *America's Munitions: 1917–1918*. Washington, DC: Government Printing Office, 1919.
55. Crozier, William. *Ordnance and the World War: A Contribution to the History of American Preparedness*. New York: Charles Scribner's Sons, 1920.
56. Culmann, Frédéric. *Tactique d'Artillerie*. Paris: Henri Charles Levrault, 1937.
57. ———. *Le Canon à Tir Rapide dans la Bataille*. Paris: Charles-Lavauzelle, 1914.

58. Dastrup, Boyd L. *King of Battle: A Branch History of the U.S. Army's Field Artillery*. Fort Monroe, VA: Office of the Command Historian, U.S. Army Training and Doctrine Command, 1992.
59. Dougan, David. *The Great Gunmaker: The Story of Lord Armstrong*. Newcastle upon Tyne, England, n.d.
60. Downey, Fairfax. *Cannonade: Great Artillery Actions of History, the Famous Cannons, and the Master Gunners*. Garden City, NY: Doubleday, 1966.
61. Dredge, James, ed. *Modern French Artillery: The St. Chamond, De Bange, Canet, and Hotchkiss Systems*. London: Offices of Engineering, 1892.
62. DuBois, Pierre Joseph Louis. *L'Artillerie de Campagne dans la Guerre Actuelle*. Paris: L. Fouchier, 1916.
63. DuCane, Sir John. *The Campaign in Alsace, August 1870*. London: Hugh Rees, 1912.
64. Dyer, Alexander B. *Handbook for Light Artillery*. New York: John Wiley and Sons, 1896.
65. Eberhard, Ritter von. *Das Wesen der modernen Visiervorrichtungen der Landartillerie*. Berlin: Verlag von A. Bath, 1909.
66. Ellis, W.B.E. *Brief Notes on Field Artillery*. London: William Clowes and Sons, 1872.
67. Elmslie, F. B. *The Possible Effects on Tactics of Recent Improvements in Weapons*. Aldershot, England: Edward Stanford, 1899.
68. Estorff, Ludwig von. *Takische Lehren aus dem Russisch-Japanischen Feldkrieg im Lichte unsrer neuesten Vorschriften*. Berlin: E. S. Mittler und Sohn, 1909.
69. Falls, Cyril B. *The Great War*. New York: G. P. Putnam, 1959.
70. Farndale, Sir Martin. *History of the Royal Regiment of Artillery, Western Front, 1914–1918*. London: Royal Artillery Institution, 1986.
71. Farrow, Edward S. *American Guns in the War with Germany*. New York: E. P. Dutton and Company, 1920.
72. Farwell, Byron. *The Great Anglo-Boer War*. New York: Harper and Row, 1976.
73. *Feld Artillerie der Zukunft*. Berlin: Friedrich Luckhardt, 1887.
74. Feline, Jacques E. P. *L'Artillerie au Maroc*. Paris: Berger-Levrault, 1902.
75. Feeser, Friedrich. *Artillerie im Feldkrieg*. Berlin: E. S. Mittler and Sohn, 1930.
76. Forbes, Arthur. *A History of Army Ordnance Service*. 3 vols. London: The Medici Society, 1929.
77. Fraser-Tytler, Neil. *Field Guns in France*. London: Hutchinson and Company Ltd., 1929.
78. Friedrich, John. *The Franco-German War*. London: Swan Sonnenschein and Company, 1900.
79. Gascouin, General. *L'Évolution de Artillerie pendant la Guerre*. Paris: E. Flammarion, 1921.
80. German General Staff. *Drill Regulations of the German Field Artillery*. London: H.M.S.O., 1893 and 1899.
81. ———. *The Evolution of Modern Field Artillery*. N.p., 1903.
82. Girardon, Étienne. *Leçons d'Artillerie, Conformes au Programme de l'École Militaire de l'Artillerie*. Paris: Berger-Levrault, 1895.
83. ———. *Organisation du Matériel d'Artillerie*. Paris: Berger-Levrault, 1903.
84. Graham, C.A.L. *History of the Indian Mountain Artillery*. Aldershot, England: Gale and Polden, 1957.

85. ———. *The Story of the Royal Regiment of Artillery*. 5th ed. Woolwich, England: Royal Artillery Institution, 1944.
86. Greene, F. V. *Report of the Russian Army and Its Campaigns in Turkey, 1877–1878*. New York: D. Appleton and Company, 1879.
87. Greener, W. W. *The Gun and Its Development*. 6th ed. New York: Charles Scribner's Sons, 1897.
88. Grierson, J. M. *Confidential Notes on the Training of German Field Artillery*. London: War Office Press, 1897.
89. Griffith, Paddy. *Forward into Battle: Fighting Tactics from Waterloo to Vietnam*. New Haven, CT: Yale University Press, 1981.
90. Gudmundsson, Bruce I. *Stormtroop Tactics: Innovation in the German Army, 1914–1918*. Westport, CT: Greenwood Press, 1989.
90A. ———. *On Artillery*. Westport, CT: Greenwood Press, 1993.
91. Guttman, Oscar. *The Manufacture of Explosives*. 2 vols. London: Whittaker, 1895.
92. Haber, L. F. *The Poisonous Cloud: Chemical Warfare in the First World War*. Oxford, England: Clarendon Press, 1986.
93. Hall, Ernest F. *Cooperation between Infantry and Artillery and the Best Means of Obtaining It*. London: Hugh Rees, 1912.
94. Hamilton, Douglas T. *High-Explosive Shell Manufacture*. New York: Industrial Press, 1916.
95. ———. *Shrapnel Shell Manufacture*. New York: Industrial Press, 1915.
96. Hamilton-Gordon, Alexander. *Changes in Field Artillery Equipment and Their Effect on the Action of Field Artillery in the Attack*. London: Hugh Rees, 1904.
97. Hamley, Edward B. *The Operations of War Explained and Illustrated*. 2nd ed. Edinburgh: William Blackwood and Sons, 1869.
98. Harvey, C. L. *The Organization and Tactical Use of Cavalry, Artillery, and Infantry*. London: Gale and Polden, 1893.
99. Henderson, G.F.R. *The Science of War*. New York: Longmans, Green, and Company, 1905.
100. Herr, Frederick G. *Artillerie*. Paris: Berger-Levrault, 1923.
101. ———. *The General Report of General Herr*. N.p., 1919.
102. Hicks, James E. *Notes on French Ordnance: 1717–1936*. Mt. Vernon, NY: James E. Hicks, 1938.
103. Hime, H.W.L. *History of the Royal Artillery*. London: Longmans, Green, and Company, 1908.
104. Hinze, Rolf. *Die Hannoverische Artillerie*. Düsseldorf: Selbstverlag des Verfassers, 1977.
105. Hoffbrauer, Ernst von. *The German Artillery in the Battles near Metz*. Translated by Captain Hollis. London: H.M.S.O., 1874.
106. ———. *Schwebende Feldartillerie Fragen: Taktischartilleristische Studie für Offiziers aller Waffen*. Berlin: E. S. Mittler und Sohn, 1904.
107. Hogg, Ian V. *The Guns: 1914–1918*. New York: Ballantine Books, 1971.
108. Hogg, O.F.G. *Artillery: Its Origin, Heyday, and Decline*. Hamden, Ct: Archon Books, 1970.
109. ———. *The Royal Arsenal: Its Background, Origin, and Subsequent History*. 2 vols. London: Oxford University Press, 1963.
110. Hohenlohe-Ingelfinger, Kraft Karlzu. *Letters on Artillery*. 2nd ed. Translated by N. L. Walford. London: Edward Stanford, 1902.

111. Homes, Robert. *A Précis of Modern Tactics*. London: H.M.S.O., 1882 and 1896.
112. Howard, Michael. *The Franco-Prussian War*. New York: Macmillan, 1961.
113. Hunt, Arthur L. *Artillery Lines of Fire*. London: F. Groom and Company Ltd., 1915.
114. Ingalls, James M. *Ballistics*. Washington, DC: Government Printing Office, 1893.
115. ———. *Interior Ballistics*. 3rd ed. New York: John Wiley and Sons, 1894.
116. Jaegle, Ernest. *L'Artillerie*. Paris: W. Hinrichsen, 1887.
117. Jobe, Joseph, ed. *Guns: An Illustrated History of Artillery*. New York: New York Graphic Society, 1971.
118. Johnson, Curt. *Artillery: The Big Guns Go to War*. London: Octopus Books, 1975.
119. Johnstone, Henry M. *A History of Tactics*. London: Hugh Rees, 1906.
120. Jones, Tom, and Fitzhugh McMaster. "Light Battery F, 5th Regiment of Artillery, Peking, 1900." *Military Collector and Historian*, Summer 1989, 86–87.
121. Kessler, Charles. *Tactique des Trois Armes*. 2nd ed. Paris: R. Chapelot, 1903.
122. Kirke, R. M. St. G. "Some Aspects of Artillery Development during the First World War on the Western Front." *Journal of the Royal Artillery*, September 1974, 130–40.
123. Klass, Gert von. *Krupps: The Story of an Industrial Empire*. London: Sidgwick and Jackson, 1954.
124. Kriegsministerium. *Drill Regulations of the German Field Artillery*. Translated by W. A. Macbean. London: H.M.S.O., 1893.
125. ———. *Drill Regulations of the German Artillery*. Translated by J.H.V. Crowe. London: H.M.S.O., 1901.
126. ———. *Entwurf zu einem Exerzier-Reglement für die Feldartillerie der königlichen Preussischen Armee*. Berlin: Vossiche Buchhandlung, 1876.
127. ———. *Gefechtsvorschrift für die Artillerie*. Berlin: Reichsdruckerei, 1917.
128. Laidley, T.T.S. *The Ordnance Department of the U.S. Army*. Washington, DC: Government Printing Office, 1874.
129. Langlois, Louis A.H. *Enseignements de Deux Guerres*. Paris: Charles-Lavauzelle, 1909.
130. ———. *Les Artilleries de Campagne de l'Europe en 1874*. Paris: Berger-Levrault, 1875.
131. ———. *L'Artillerie de Campagne en Liaison avec les Autres Armes*. 3 vols. Paris: R. Chapelot, 1892 and 1908.
132. ———. *Lessons from Two Recent Wars*. London: War Press Office, 1909.
133. LeBlond, Henri L. *Le Canon à Tir Rapide et l'Instruction de l'Artillerie*. Paris: Charles-Lavauzelle, 1906.
134. Liddell Hart, B. H. *The Real War: 1914–1918*. Boston: Little, Brown, 1930.
135. Lissak, Ormond. *Ordnance and Gunnery*. New York: John Wiley and Sons, 1907.
136. Lloyd, E. W., and A. G. Hadcock. *Artillery: Its Progress and Present Position*. Portsmouth, England: J. Griffin and Company, 1893.
137. Longridge, James A. *The Artillery of the Future and the New Powers*. London: E. and F. N. Spon, 1891.
138. Lupfer, Timothy T. *The Dynamics of Doctrine: The Changes in German Tactical Doctrine during the First World War*. Fort Leavenworth, KS: Combat Studies Institute, U.S. Army Command and General Staff College, 1981.
139. McNeill, William H. *The Pursuit of Power: Technology, Armed Forces, and Society since* A.D. *1000*. Chicago: University of Chicago Press, 1982.

140. Manchester, William. *The Arms of Krupp, 1587–1968*. Boston: Little, Brown, 1968.
141. May, Sir Edward S. *A Retrospect on the South African War*. London: Sampson, Low, Marston, and Company, 1901.
142. ———. *Achievements of Field Artillery*. Woolwich, England: Royal Artillery Institution, 1893.
143. ———. *Field Artillery with the Other Arms: Its Employment, Illustrated from Military History, and Its Re-armament with Quick-Firing Guns Discussed*. London: S. Low, Marston and Company, 1898.
144. Menne, Bernhard. *Krupp; or, The Lords of Essen*. London: Hodge, 1937.
145. Metcalfe, Henry. *A Course of Instruction in Ordnance and Gunnery*. New York: John Wiley and Sons, 1894.
146. Millis, Walter. *The Martial Spirit*. Boston: Houghton Mifflin, 1931.
147. Ministère de la Guerre. *Cours: Special à l'Usage des Sous-Officiers d'Artillerie*. Paris: L. Baudoin, 1886.
148. Monthaye, E. *Krupp and de Bange*. 2nd ed. Translated by O. E. Michaelis. New York: T. Prosser and Son, 1888.
149. Montu, Carlo. *Stòria dello Artigilierìa Italiana*. 13 vols. Rome: Edita a Cura della Rivista d'Artiglierìa e Gènio, n.d.
150. Moretti, Onorio. *Notes on Training Field Artillery Drill*. New Haven, CT: Yale University Press, 1917.
151. Morgan, Frederick C. *Handbook of Artillery Materiel*. 6th ed. London: Clowes and Sons, 1899.
152. Morgan, H. Wayne. *America's Road to Empire: The War with Spain and Overseas Expansion*. New York: John Wiley and Sons, 1965.
153. Morrison, Charles C. *Modern Guns and Mortars*. Washington, DC: Government Printing Office, 1893.
154. Morrison, Charles C., and James C. Ayres. *Course of Instruction for Artillery Gunners: Modern Guns and Mortars Adopted in the United States Land Service, Their Carriages, Projectiles, Fuzes, and Sights*. Washington, DC: Government Printing Office, 1893.
155. Nash, David. *German Artillery, 1914–1918*. London: Almark Publishing Company Ltd., 1970.
156. Nesmith, Vardell E. "The Quiet Paradigm Change: The Evolution of the Field Artillery Doctrine of the United States Army, 1861–1905." Doctoral dissertation, Duke University, 1977.
157. Newbold, J. T. *How Europe Armed for War*. London: Blackfriars Press, 1916.
158. Nicholls, D. F. *Our Artillery as Used in the Present War in South Africa and concerning Heavy Guns and Mountains in Particular*. Malta: Malta Tactical Archaeological and Scientific Society, 1902.
159. Nicholson, G.W.L. *The Gunners of Canada: The History of the Royal Regiment of Canadian Artillery*. Toronto: McClelland and Stewart, 1967.
160. Nicholson, W. A. *Artillery Fire: The Battery*. London: Edward Stanford, 1907.
161. Niessel, Henri A. *Combinaison des Efforts de l'Infanterie et de l'Artillerie dans le Combat*. Paris: Charles-Lavauzelle, 1908.
162. Nobles, Sir Andrew. *Artillery and Explosives: Essays and Lectures Written and Delivered at Various Times*. London: John Murray, 1906.
163. Norton, Charles B., and W. J. Valentine. *Report to the Government of the United*

States on the Munitions of War Exhibited at the Paris Universal Exhibition, 1867. Washington, DC: Government Printing Press, 1868.

164. Ogorkiewicz, R. M. "The Evolution of Self-propelled Guns." *Military Review*, October 1951, 102–9.
165. Owens, Charles H. *The Principles and Practice of Modern Gunnery, Including Artillery Materiel, Gunnery, and Organization and Use of Artillery in Warfare*. London: John Murray, 1871.
166. Pakenham, Thomas. *The Boer War*. New York: Random House, 1979.
167. Paloque, Henri J. *Artillerie de Campagne*. Paris: Octave Doin et Fils, 1909.
168. Pardieu, Major de. *A Critical Study of German Tactics and of the New German Regulations*. Translated by Charles F. Martin. Fort Leavenworth, KS: U.S. Cavalry Association, 1912.
169. Pemberton, A. L. *The Development of Artillery Tactics and Equipment*. London: The War Office, 1950.
170. Percin, Alexandre. *Artillery at the Picardy Maneuvers in 1910*. Translated by the General Staff. London: H.M.S.O., 1912.
171. ———. *Cinq Années de Inspection*. Paris: Chapelot, 1912.
172. Plessix, Hipployte I. *Manuel Complète d'Artillerie*. Paris: L. Baudoin, 1883.
173. Pratt, Sisson C. *Field Artillery: Its Equipment, Organization, and Tactics*. London: K. Paul, Trench, and Company, 1883.
174. Rimailho, François. *Artillerie de Campagne*. Paris: Gauther Villars, 1924.
175. Roberts, Joseph. *The Handbook of Artillery for the Service of the United States Army and Militia*. New York: D. Van Nostrand, 1875.
176. Roberts, Larry D. "The Artillery with the Regular Army in the West." Doctoral dissertation, Oklahoma State University, 1981.
177. Rohne, Heinrich. *An Essay on the Shrapnel Fire of Field Artillery*. Translated by Colonel N. L. Walford. London: Royal Artillery Institution, 1896.
178. ———. *L'Artillerie de Campagne Française*. Paris: Berger-Levrault, 1903.
179. ———. *Progress of Modern Field Artillery*. Translated by Montgomery M. Macomb. Washington, DC: Gibson Brothers, 1908.
180. Romain, Charles A. *Les Responsabilités de l'Artillerie Française en 1870*. Paris: Berger-Levrault, 1913.
181. Roskaten, Captain von. *Die heutige Feldartillerie mit Rohrrücklauf: Ihr Materiel, technische Hilfsmittel, Schiessverfahren, Organisation, und Taktik*. Berlin: R. Eisenschmidt, 1909.
182. Rouquerol, Jean G. M. *L'Artillerie dans le Bataille du 18 Aout*. Paris: Berger-Levrault, 1906.
183. ———. *The Tactical Employment of Quick Firing Field Artillery*. Translated by P. Radcliffe. London: Hugh Rees, 1903.
184. Rouvier, Jacques. *Present-Day Warfare: How an Army Trains and Fights*. New York: Charles Scribner's Sons, 1918.
185. Roy, Joseph A. *Historie de la Famille Schneider et du Creusot*. Paris, 1962.
186. Russell, H. T. *The Employment of Artillery in the Field in Co-operation with the Other Arms*. London: Hugh Rees, 1902.
187. Savoy, Wallace J. "The Evolution of the American Modern Light Field Gun." Master's thesis, U.S. Army Command and General Staff College, 1978.
188. Scales, Robert H. "Artillery in Small Wars: The Evolution of British Artillery Doctrine, 1860–1914." Doctoral dissertation, Duke University, 1976.

189. Schell, Adolf von. *The Tactics of Field Artillery*. Translated by A. E. Turner. London: H.M.S.O., 1889.
190. Schloesing, M. Th. *The French "75"*. Translated by Frank H. Canaday. Menasha, WI: George Banta Publishing Company, 1918.
191. Scott, J. D. *Vickers*. London, 1962.
192. Secretary of State for War, Great Britain. *Treatise on the Manufacture of Guns and Text Book of Ordnance in the British Service*. London: H.M.S.O., 1887.
193. Snow, William J. *History of the Office, Chief of Field Artillery*. Washington, DC: n.p., 1933.
194. ———. *Signposts of Experience: World War Memoirs of Major General William J. Snow*. Washington, DC: U.S. Army Field Artillery Association, 1941.
195. Spaulding, Oliver L. *Notes on Field Artillery for Officers of All Arms*. Fort Leavenworth, KS: U.S. Cavalry Association, 1907.
196. ———. *Notes on Field Artillery*. Leavenworth, KS: U.S. Cavalry Association, 1908.
197. ———. *Weapons and Munitions of War: Artillery Weapons*. Fort Leavenworth, KS: Infantry and Cavalry School, 1907.
198. Stokesbury, James L. *A Short History of World War I*. New York: William Morrow, 1981.
199. Stoney, Francis. *A Text Book of the Construction and Manufacture of the Rifled Ordnance in the British Service*. London: H.M.S.O., 1872.
200. Sulzbach, Herbert. *With the German Guns: Four Years on the Western Front, 1914–1918*. Hamden, CT: Archon Books, 1981.
201. The War Office, Great Britain. *Field Artillery Drill*. London: H.M.S.O., 1889, 1893, and 1896.
202. ———. *Field Artillery Service Handbook*. London: H.M.S.O., 1897.
203. ———. *Field Artillery Training*. London: H.M.S.O., 1902, 1906, 1912, and 1914.
204. ———. *Instructions for Practice, Field Artillery*. London: H.M.S.O., 1889–1914.
205. ———. *Manual of Artillery Exercises*. London: H.M.S.O., 1873, 1877, 1879.
206. ———. *Military History of the Campaign of 1882 in Egypt*. London: H.M.S.O., 1908.
207. Trask, David F. *The War with Spain in 1898*. New York: Macmillan, 1981.
208. Trebilcock, Clive A. *The Vickers Brothers: Armaments and Enterprise, 1854–1914*. London: Europa Publishers, 1977.
209. Trequier, François A. *L'Artillerie pour les Autres Armes*. 2nd ed. Paris: Charles-Lavauzelle, 1913.
210. Tschappat, William H. *Textbook of Ordnance and Gunnery*. New York: John Wiley and Sons, 1917.
211. Ubique. *Famous Battles of the Royal Artillery*. Portsmouth, England: W. H. Barrell Ltd., 1930.
212. U.S. Army, American Expeditionary Force. *Notes on German Artillery Materiel*. General Headquarters, American Expeditionary Force, 1918.
213. U.S. Army Field Artillery School. *History of the Development of Field Artillery Materiel*. Fort Sill, OK: Field Artillery School, 1939.
214. ———. *Materiel*. Fort Sill, OK: Field Artillery School, 1932 and 1937.
215. U.S. Army War College. *Field Artillery Notes*. Washington, DC: Government Printing Office, 1917.
216. U.S. Ordnance Department. *Collection of Annual Reports and Other Important*

Papers Relating to the Ordnance Department: 1812–1889. Washington, DC: Government Printing Office, 1812–1889.

217. ———. *Handbook of Ordnance Data*. Washington, DC: Government Printing Office, 1918.
218. Verner, Rudolf H. C. *Guns and Projectiles*. 2nd ed. London: John Hogg, 1915.
219. Wade, Aubrey. *The War of the Guns: Western Front, 1917 and 1918*. New York: Charles Scribner's Sons, 1936.
220. Wagner, Arthur L. *Organization and Tactics*. 2nd ed. Kansas City, MO: Hudson-Kimberly Publishing Company, 1894.
221. War Department (U.S.). *Artillery Tactics*. New York: D. Appleton and Company, 1888 and 1889.
222. ———. *Drill Regulations for Field Artillery, U.S. Army (Provisional)*. Washington, DC: Government Printing Office, 1905.
223. ———. *Drill Regulations for Field Artillery, U.S. Army (Provisional)*. Washington, DC: Government Printing Office, 1907.
224. ———. *Drill Regulations for Field Artillery, U.S. Army*. Washington, DC: Government Printing Office, 1908.
225. ———. *Drill Regulations for Field Artillery (Horse and Light)*. Washington, DC: Government Printing Office, 1911.
226. ———. *Drill Regulations for Light Artillery*. Washington, DC: Government Printing Office, 1896.
227. ———. *Instruction for Field Artillery*. Washington, DC: Government Printing Office, 1867.
228. ———. *Light Artillery Drill Regulations*. Washington, DC: Government Printing Office, 1891.
229. ———. *Modern Guns and Gunnery: A Practical Manual for Officers of Horse, Field, and Mountain Artillery*. Washington, DC: Government Printing Office, 1907.
230. ———. *Provisional Drill and Service Regulations for Field Artillery (Horse and Light)*. Washington, DC: Government Printing Office, 1916 and 1917.
231. Warner, Denis A. *The Tide at Sunrise: A History of the Russo-Japanese War*. New York: Charterhouse, 1974.
232. Wernigk, von. *Taschenbuch für die Feldartillerie*. Berlin: E. S. Mittler und Sohn, 1886–1908.
233. Westervelt, William I. *Gunnery and Explosives for Field Artillery Officers*. Washington, DC: Government Printing Office, 1911.
234. Wheeler, Charles B. *A Discussion of the Proposed Methods to Increase the Rapidity of Fire of Field Guns*. Washington, DC: Government Printing Office, 1901.
235. Wille, R. *Granatschrapnell-Brisanzschrapnell*. Berlin: R. Eisenschmidt, 1911.
236. Wisse, Sébastien Léon. *A la Gloire de l'Artillerie Française de 1870s à 1918*. Paris: Charles Lavauzelle, 1903.
237. Wolseley, Garnet. *The Soldier's Pocketbook*. London: Macmillan, 1887.

8

From Peace to Cold War, 1919–1991

GENERAL HISTORIES

To complement the general studies identified in the first bibliographical essay see Allan R. Millett and Williamson Murray's three-volume *Military Effectiveness* (173). Their edited work provides a systematic, comprehensive survey of the military effectiveness of France, Germany, the Soviet Union, the United States, Great Britain, Japan, and Italy from 1914 to 1945. Contributing authors examine tactics, strategy, politics, and operations.

Although the literature on World War II is voluminous, several general works are worth examining for a start. James L. Stokesbury's *A Short History of World War II* (237) supplies a useful overview with a solid bibliography. H. P. Willmott's *The Great Crusade: A New Complete History of the Second World War* (279) narrates the entirety of World War II. He covers operations, politics, strategy, and tactics and lays a solid foundation for future analyses of the war. Other books on World War II are V. J. Esposito's *A Concise History of World War II* (73), L. L. Snyder's *The War: A Concise History, 1939–1945* (233), and B. H. Liddell Hart's *History of the Second World War* (161). For official histories of combatants written through the late 1960s see Robin Higham's (ed.) *Official Histories: Essays and Bibliographies from Around the World* (125).

For a general account of war since 1945 see Michael Carver's *War since 1945* (41). He surveys British and French colonial conflicts, American ventures, and conventional clashes. Martin Van Creveld's *The Transformation of War* (266) is a provocative examination of war since World War II. The author analyzes the role of guerrillas and terrorists in changing warfare. Books on the Korean War include Roy E. Appleman's *South to the Naktong, North to the Yalu* (7), which is one of the best accounts of the early phases of the war, Robert Leckie's

Conflict: The History of the Korean War (157), David Ree's *Korea: The Limited War* (204), and T. R. Fehrenbach's *This Kind of War: A Study in Unpreparedness* (79).

For the Vietnam War consult Bernard Falls's *Street without Joy* (76), Stanley Karnow's *Vietnam: A History* (151), Larry Berman's *Planning a Tragedy: The Americanization of the War in Vietnam* (23), George C. Herring's *America's Longest War: The United States and Vietnam, 1950–1975* (123), David R. Palmer's *Summons of the Trumpet* (188), and Guenther Lewy's *America in Vietnam* (159).

Numerous works on the wars in the Middle East exist. One of the best is David Kimche and Dan Bawly's *The Six-Day War: Prologue and Aftermath* (153). Another outstanding study is Chaim Herzog's *The Arab-Israeli Wars: War and Peace in the Middle East from the War of Independence through Lebanon* (124). Anthony H. Cordesman and Abraham R. Wagner's three-volume *The Lessons of Modern War* (51) is a solid analysis of the Arab-Israeli conflicts from 1973 to 1989 and the Iran-Iraq war of the 1980s.

Even though participants in the Persian Gulf War of early 1991 are still drawing lessons from combat, books are beginning to appear. One of the best is Paddy Griffith's *The Ultimate Weaponry* (112). Griffith provides a penetrating discussion on the impact of the new military technology upon the Gulf War and supports his narrative with charts, photographs, and illustrations. Additional works include Martin Yant's *Desert Mirage: The True Story of the Gulf War* (282), Eliot Brenner and William Harwood's *Desert Storm: The Weapons of War* (34), James F. Dunnigan and Austin Bay's *From Shield to Storm: High Tech Weapons, Military Strategy, and Coalition Warfare in the Persian Gulf* (68), Dennis Menos's *Arms over Diplomacy: Reflections on the Persian Gulf War* (170A), and Norman Friedman's *Desert Victory: The War for Kuwait* (91). A solid article on the Gulf War is Rupert Smith's "The Gulf War: The Land Battle" (231). For a fine bibliography examine A. Sue Goodman's (ed.) *Persian Gulf War, 1990–1991* (101A), which is part of the Air University Library's Special Bibliography Series.

FIELD ARTILLERY HISTORIES

Histories of field artillery since World War I are limited. One of the best is Frank E. Comparato's *Age of Great Guns: Cannon Kings and Cannoneers Who Forged the Firepower of Artillery* (46). Jonathan B.A. Bailey's *Field Artillery and Firepower* (12) is the best source available on field artillery technology, organization, and tactics during the years since 1919 and has an excellent bibliography of journal articles. Both Comparato's and Bailey's books are excellent starting points for studying field artillery in the twentieth century.

Bruce I. Gudmundsson's *On Artillery* (112A) deals with field artillery warfare in the twentieth century, both pre- and post-1919. This book makes use of extensive German and French sources, and discusses technical issues in non-technical language.

For more see O.F.G. Hogg's *Artillery: Its Origin, Heyday, and Decline* (135) for its solid coverage of field artillery developments since 1919. Another book is Phillip H. Stevens's *Artillery through the Ages* (236). In *A History of Artillery* (215) H.C.B. Rogers not only furnishes the reader with a general survey of artillery history but also includes chapters on post–World War I field artillery. Joseph Jobe's (ed.) *Guns: An Illustrated History of Artillery* (148) has a short section on field artillery since World War I. Dudley Pope's *Guns* (197) is also worth surveying. John Batchelor and Ian V. Hogg's illustrated *Artillery* (15) should be considered for its thoughtful examination of self-propelled artillery. Curt Johnson's *Artillery: The Big Guns Go to War* (149) is an overview of field artillery history since the mid-eighteenth century, but it also covers rocket and self-propelled artillery. Ian V. Hogg's *The Illustrated Encyclopedia of Artillery* (134) is another invaluable source. V. Drake-Brockman's "Modern Developments in Foreign Artilleries of the Field Army" (67) is a short history of field artillery.

Ian V. Hogg's *Artillery in Color: 1920–1963* (130) is useful, as is his *The Guns: 1939–1945* (133). The latter is an illustrated narrative of the field guns employed during World War II. Fairfax Downey's *Cannonade: Great Artillery Actions of History, the Famous Cannons, and the Master Gunners* (65) has moving accounts of operations in World War II as well as the Korean War. Peter Chamberlain's illustrated *Heavy Artillery: World War II Fact File* (42) is an excellent reference book on World War II technology. The same author's illustrated *Light and Medium Field Artillery: World War II Fact File* (43) complements his book on heavy artillery. For a general overview of field artillery during World War II see G. Tauson's "Artillery Development in World War II" (242). Robert Scale's *Firepower in Limited War* (221) covers tactics, doctrine, and field artillery in small wars, especially since 1945. Franz Kosar's *A Pocket History of Artillery: Light Field Guns* (154) is also worthwhile.

Jane's Armour and Artillery (145) is invaluable for field artillery technology of the 1970s, as is Christopher F. Foss's *Artillery of the World* (86). For field artillery in the 1980s the ten-volume set of *Jane's Armour and Artillery* (146) and the ten-volume set of *Jane's Weapons Systems* (147) are vital for their short histories of weapons and charts of weapon characteristics. Christopher F. Foss's *Jane's Pocket Book of Towed Artillery* (87) furnishes a broad coverage of worldwide field artillery systems, technical specifications, and pictures. Shelford Bidwell's (ed.) *Brassey's Artillery of the World* (29) is critical, as is Terry J. Gander's *Artillery* (94). The most recent book on field artillery is Ian V. Hogg's *Artillery 2000* (129). It provides a general overview of field artillery materiel, ammunition, and employment as they existed at the close of the 1980s, has histories of the major weapon systems, and examines systems scheduled to be introduced in the near future.

For the Persian Gulf War of early 1991, military journals are particularly valuable sources. Representative articles include L. Scott Lingamfelter's "In the

Wake of the Storm: Improving the FA [Field Artillery] after Operation Desert Storm'' (162) and Michael D. Holthus and Steven M. Chandler's ''Myths and Lessons of Iraqi Artillery'' (136). Other articles in *Journal of the Royal Artillery* and *Field Artillery* are available for their insights into Allied artillery as well as their critique of Iraqi artillery.

Literature on field artillery of a specific country should also be considered. A valuable book on German artillery is Ian V. Hogg's *German Artillery of World War II* (132). Another one is Donald B. McClean's *Illustrated Arsenal of the Third Reich* (166). A. G. Cole's ''German Artillery Concentrations in World War II'' (44) is brief but excellent. Leslie W. Barnes's *Canada's Guns: An Illustrated History of Artillery* (14) is a short history of Canadian artillery. G.W.L. Nicholson's *The Gunners of Canada: The History of the Royal Regiment of Canadian Artillery* (178) is especially valuable for Canadian field artillery.

American field artillery is well covered in Boyd L. Dastrup's *King of Battle: A Branch History of the U.S. Army's Field Artillery* (55), which should be the basis for studying American field artillery during the era under consideration. Fairfax Downey's *Sound of the Guns: The Story of American Artillery* (66) contains interesting chapters on field artillery in World War II and Korea and on guided missiles. The U.S. Army Field Artillery School's *Artillery in Combat* (254) provides interesting insights into American field artillery operations during World War II. Ian V. Hogg's *British and American Artillery of World War II* (131) should be considered for field artillery of the 1940s.

One of the fathers of American organic aerial observation for field artillery, William W. Ford, discusses the fight to implement organic air observation in the U.S. Army in *Wagon Soldiers* (84). The same author's ''Grasshoppers'' (83) also has indispensable insights into the evolution of organic air observation in the U.S. Army, while Ken Wakefield's *The Fighting Grasshoppers: U.S. Liaison Aircraft Operations in Europe, 1942–1945* (269) is a nicely illustrated popular account of the origins and growth of U.S. liaison aircraft for aerial observation and other missions during World War II. C. H. Donnelly's ''The Artillery Aerial OP'' (61) covers organic aerial observation in the U.S. Army during the first years of American participation in World War II. For a more analytical discussion of air observation see Laurence B. Epstein's ''Army Organic Light Aviation: The Founding Fathers'' (70). *A History of Army Aviation: 1950–1962* (275), written by Richard P. Weinert, Jr., has a solid chapter on early aviation in the U.S. Army that covers aerial observation.

In *The Cannon's Mouth: The Role of U.S. Artillery during World War II* (105) Lewis J. Gorin, Jr., follows the operations in Europe of the 6th Field Artillery Group and one of its battalions, the 59th Armored Field Artillery Battalion. Frank G. Ratliff's ''The Field Artillery Fire Direction Center: Its Past, Present, and Future'' (202) is a key article on the origin of the fire direction center in the U.S. Army during the 1930s. For a solid study of the U.S. Army's field artillery group for nondivisional artillery during World War II through the Korean War consult Russell A. Weathersby's ''The Field Artillery Group in Support of the Corps and Field Army, 1942–1953'' (274). Another essential discussion of the role of field artillery groups in the U.S. Army during World War II is J. A. Crane's ''Field Artillery Groups'' (52).

For more on American field artillery in Europe during World War II see C. E. Hart's ''Artillery with an American Army in Europe'' (118). Another article on American artillery in World War II is J. R. Reeves's ''Artillery in the Ardennes'' (205). Larry D. Gahagan's ''Development and Use of Field Artillery Proximity Fuzes in World War II'' (93) furnishes an excellent discussion of the introduction of the proximity fuze in the U.S. Army. For a useful discussion of a much-neglected topic see William R. Bursell's ''American Sound Ranging in Four Wars'' (38). Although the article is short, it is worth reading.

For an overview of American tactics since World War II consult Robert A. Doughty's *The Evolution of U.S. Army Tactical Doctrine, 1946–1976* (64). He places field artillery tactics and operations in the context of combined arms warfare as it evolved during the thirty years following World War II. American artillery in the Korean War is covered in Russell A. Gugeler's *Combat Actions in Korea: Infantry, Artillery, Armor* (113).

David Ott's *Field Artillery: 1954–1973* (186) is the standard source for American field artillery in the Vietnam War, although it lacks footnotes and a bibliography. However, a draft copy of the book with footnotes and bibliography is available at the Morris Swett Technical Library, U.S. Army Field Artillery School, Fort Sill, Oklahoma. James R. Arnold's popular account, *The Illustrated History of Artillery: The Vietnam War* (8), tells a wonderful story but lacks footnotes and a bibliography to substantiate facts. Another book on American field artillery in the Vietnam War is Jim Gleckler's *Redleg* (101).

Although J. W. Ryan's *Guns, Mortars, and Rockets* (218) is a textbook on technology, it still gives the reader a picture of American and British field artillery of the late 1970s and early 1980s. Steven J. Zaloga and Arnold Meisner's *U.S. Mechanized Firepower* (284) is especially valuable for self-propelled artillery of the 1980s.

Bibliographical reference books and lineage and honors books also exist for American field artillery. A most valuable book is Ralph W. Donnelly's *An Annotated Bibliography of the United States Marine Corps' Artillery* (62), while Janice A. McKenney's *Field Artillery* (168) is critical for lineage and honors data and short bibliographies on U.S. Army field artillery regiments. James A. Sawicki's *Field Artillery Battalions of the U.S. Army* (220) is lineage and honors book for U.S. Army field artillery battalions that saw service in World War II.

For a solid examination of the history of self-propelled artillery in the British army consult Philip Ventham and David Fletcher's *Moving the Guns: The Mechanization of Royal Artillery* (267). They trace the development of tanks as well as artillery since 1846 and cover World War I and the interwar years. Shelford Bidwell's *Gunners at War: A Tactical Study of the Royal Artillery in the Twentieth Century* (27) is a solid study. Although it is short, H. Rowan-Robinson's *Artillery: Today and Tomorrow* (217) analyzes British field artillery materiel, organization, and tactics in the 1920s. Shelford Bidwell and Dominick Graham's *Firepower: British Army Weapons and Theories of War, 1904–1945* (30), with its solid bibliography, is an outstanding study of technology, tactics, and organization. Kenneth Brookes's *Battle Thunder: The Story of Britain's Artillery* (35) provides an excellent background on British field artillery. Another essential book on British field artillery is A. L. Pemberton's *The Development of Artillery Tactics and Equipment* (192). C.A.L. Graham's *The History of the Indian Mountain Artillery* (109) is also worth reading. For organic aerial observation read H. J. Parham's *Unmanned into Battle: The Story of the Air Observation Post* (189).

R.G.S. Bidwell's "The Development of British Field Artillery Tactics, 1920–1939" (24), "The Development of British Field Artillery Tactics, 1940–1942" (25), and "The Development of British Field Artillery Tactics, 1940–1943" (26) are indispensable for understanding British field artillery tactics during World War II. In "The Royal Artillery of the British Army" (56), Norman Dodd provides an overview of British field artillery history from the founding of the Royal Artillery Regiment in 1716 through the 1970s. "The Queen's Gunners: British Field Artillery" (245) discusses British field artillery technology and organization at the beginning of the 1960s.

Any study of Soviet field artillery must begin with Christopher Bellamy's *Red God of War: Soviet Artillery and Rocket Forces* (21) for its thoughtful interpretation of artillery organization, tactics, and weapons, and extensive bibliography of Russian- and English-language sources. Frederick P. A. Hammersen's unpublished master's thesis, "A History of Russian and Soviet Artillery, 1914–1941" (116), is useful for Soviet field artillery and has an excellent Russian- and English-language bibliography. For more on Russian and Soviet field artillery see the U.S. Army Command and General Staff College's *Organization and Employment of a Division, Soviet (Russian) Army* (251). The War Department (U.S.) Military Intelligence Service's *The Organization and Tactical Employment of Soviet Ground Forces* (272) should be examined. Interestingly, the title is translated from Russian, but the text is not. For more on Soviet tactics consult I. S. Prochko's *Artillery: God of War* (200) and V. D. Grendall's *Artillery in Principal Roles of Combat* (111). H. G. de Watteville's "Russian Artillery: 1941–1945" (273) provides invaluable insights into Russian field artillery operations, tactics, and technology during World War II. V. Smirnov's "Soviet Artillery in World War II" (229) is excellent for Soviet field artillery.

Christopher Bellamy's "Soviet Artillery and Rocket Forces: 1940–1980" (22) is an outstanding discussion of Soviet artillery technology. Another sound article on Soviet field artillery is John Gordon's "The Evolution of Soviet Fire Support, 1940–1988" (104). The author examines organization, doctrine, and materiel. For a solid scrutiny of Soviet self-propelled artillery read Andrew W. Hull's "Evolution of Soviet Self-propelled Artillery" (137). D. M. Hart and E. M. Gormley's "The Evolution of Soviet Interest in Atomic Artillery" (119) is interesting. Another invaluable article is K. Hoffman's "An Analysis of Soviet

Artillery Development'' (128). It is especially useful for Soviet field artillery developments in the 1970s.

TECHNOLOGY

For field artillery technology see Thomas Hayes's *Elements of Ordnance* (121). James P. Kelly's (ed.) *Field Artillery Materiel* (152) is invaluable for field artillery technology at the beginning of the 1920s. Earl McFarland's *Textbook of Ordnance and Gunnery* (167) is another vital source. John E. Burchard's (ed.) *Rockets, Guns, and Targets* (37) covers the modernization of European and American field artillery in the 1930s. R. M. Ogorkiewicz's ''The Evolution of Self-propelled Guns'' (181) is worth reading for its insights on early self-propelled artillery. Fred K. Vigman's ''The Theoretical Evaluation of Artillery after World War I'' (268) should be examined for its critique of weapon capabilities.

For a general examination of rocketry see George E. Pendray's *The Coming of the Age of Rocket Power* (193) and M. L. O'Hagen's *Missiles and the Revolution in Warfare* (183). Both books cover the rise of rockets as field artillery weapons. A. R. Weyl's *Guided Missiles* (277) provides a useful analysis of early guided missile technology. For more on rocket technology consult I. J. Harrell's ''Development and Employment of Rockets'' (117), which is about World War II rocketry. A. Kamarck's ''Rebirth of the War Rocket'' (150) analyzes the rationale for field artillery rockets and their employment. ''Allied Artillery Missiles'' (3) discusses European rocket and missile artillery in the 1950s.

For almost fifteen years following World War II most European countries lacked the resources to modernize their field artillery because of the extensive efforts to rebuild war-torn economies, and, therefore, devoted little attention to rearming with new weapons. Rearmament picked up in the 1960s as the economies became more prosperous. During that decade and succeeding ones, many European armies and the American army initiated extensive modernization efforts to replace worn-out field artillery weapons that had been around for years, some of them since World War II. For an overview see Walter Stone's ''Current Trends in Artillery'' (238). In ''Yom Kippur and the Artillery'' (247) W.F.K. Thompson explores the impact of the war upon the field artillery. Although it is short, Christopher Campbell's *Nuclear Weapons Fact Book* (39) is informative on weapons of the 1980s.

Michael J. Crutchley and John H. Milam's ''The Modernisation of NATO's Tube Artillery'' (54) is a solid examination of endeavors to introduce new field artillery that would be effective into the twenty-first century. ''New Generation Weapons and the Doctrine of Tactics'' (177) gives the reader an overview of field artillery materiel during the first years of the 1980s. Hobart B. Pillsbury's ''Artillery: New Developments'' (196) discusses the reasoning behind the new field artillery developments that were beginning to appear early in the 1980s. ''Considerations on the Modern Tube Artillery'' (49) is invaluable for its insights into modernization, as is P. D. Williams's ''Space-Age Artillery'' (278) because of its look at future field artillery.

In ''Artillery on the Battlefield: The Future of Towed Guns'' (74) I. D. Evans challenges the utility of towed field artillery in combat in Europe. For a good analysis of the A–70, a towed 155mm howitzer, see R.C.F. Craven's ''A70 for the 90s'' (53). The author covers A–70 technology, tactics, and organization. R.D.M. Furlong's ''Self-Propelled 155mm Artillery for the 90s'' (92) examines modernization ventures in Europe and the United States to introduce new weapons and improve old ones. One should also read ''Self-Propelled Artillery for the Late 1980s'' (184) by T. J. O'Malley for a look into the future of self-propelled field artillery from the vantage point of 1985. ''Concept for Self-propelled Howitzer'' (48) is useful for its look at future field artillery technology. Ngo Niemzig's ''The SP–70–1'' (179) is invaluable for its insights into the reasons for developing the SP–70 self-propelled 155mm howitzer.

Key articles on the computerization of fire direction exist. Matthew J. Ringer and Martell D. Fritz's ''FADAC Computations versus Manual Computations'' (209) is an invaluable article on one of the first computers employed for calculating fire direction. ''TACFIRE: An Automated Artillery Fire Direction System'' (241) is an essential article on the U.S. Army's second-generation computer.

For British computer technology see ''BATES: A Layman's Guide'' (16) and M. L. O'Hagan's ''BATES: A Layman's Guide'' (182). ''Artillery Command and Control'' (9) is also worth examining. ''Computers and Artillery: From FACE to BATES'' (47) is excellent. N. F. Evan's ''The Impact of BATES on the Royal Artillery'' (75) discusses the introduction of automated data processing in the British army's field artillery. For a general projection of the role of computers in the field artillery consult A. J. Gray's ''Automatic Data Processing in the Field Artillery in the 1980s'' (110).

Literature on precision-guided munitions is also available for understanding the improvement in field artillery lethality. D. W. Hoffman's ''Developments in Precision Guided Munitions'' (127) provides valuable coverage of these munitions during the late 1970s and early 1980s. Paul F. Walker's ''Precision Guided Weapons'' (270) is useful for smart munitions developed since the Yom Kippur War of October 1973. Christopher F. Foss's ''Munitions Development'' (88) is must reading on smart munitions in the mid–1980s. Also see J. C. Antoniotti's ''Precision Guided Munitions'' (6) for an overview of the smart munitions and their employment.

In ''Precision Guided Artillery: First and Second Generation Projectiles'' (115) James F. Hall examines the rationale for precision-guided munitions and discusses the American Copperhead projectile. Another solid article defending Copperhead is ''The Copperhead'' (244). T. J. Jackson's ''Precision Guided Munitions for the U.S. Army'' (143) reviews the initial developments in precision-guided munitions, such as the Copperhead, and their dependence upon laser designators to hit a target.

In ''U.S. Tube-Launched Smart Anti-Tank Munitions'' (239) Gowri S. Sundaram covers the initial developments in the U.S. Army's program to acquire precision-guided munitions as a means to offset Soviet and Warsaw Pact numerical superiority and to reduce the number of artillery rounds necessary to disable or kill armor. Bill Rittenhouse's ''MLRS: Smart Munitions'' (211) is an excellent article on precision-guided projectiles.

John Reed's ''Beyond Copperhead: What Prospects for SADARM and STAFF?'' (203) is invaluable for the author's discussion of munitions that do not depend on laser designators for hitting targets. ''SADARM Scores Twice'' (219) is a must for understanding the implications of fire-and-forget munitions and their lethality. P. Seidenmann and D. J. Spanovich's ''SADARM'' (226) explains smart munitions technology in the U.S. Army designed to replace the Copperhead and other laser-designated munitions. For a German perspective on precision-guided munitions see Wolfgang Flume's ''Advanced Artillery Munitions: A German View'' (81). Bohdan Prehar's ''Artillery Scatterable Mines'' (198) is a short discussion of a new field artillery munition.

Unlike the others, A. K. Lorber's ''Active Defense against Precision Guided Munitions for the Future Battlefield'' (164) examines possible defenses against smart munitions. I. K. Forward's ''The Case against Copperhead'' (85) questions the use of Copperhead.

Articles on remotely piloted vehicles for target acquisition are invaluable. For a broad look consult Tim Wrixon's ''RPV Development Accelerates'' (281). Bill Sweetman's ''Unmanned Air Vehicles Make a Comeback'' (240) explains the technology and the logic behind employing such vehicles. W. R. Lippincott's ''RPV: Tomorrow's Armchair Giant Killer'' (163) projects the future of unmanned aerial vehicles on the battlefield and discusses the state of the technology. Philip J. Mills's ''RPVs over the Bekaa Valley'' (174) explores the role of unmanned aerial vehicles in the Israeli military. Irwin Stambler's ''U.S. Remotely Piloted Vehicle Program'' (234) probes the initial American work on unmanned aerial vehicles for target acquisition and real-time intelligence. For a discussion of the troubles associated with remotely piloted vehicles see D. L. Harvey's ''A Troubled Nest for Aquila'' (120). For German work with unmanned aerial vehicles consult ''German RPV Programs'' (99).

For a short study of the resurgence in the interest in multiple rocket launchers during the 1970s and 1980s see D.W.L. Robinson's ''Why Rockets?'' (214). ''Artillery Rocket Systems: A Unique Market'' (10) covers the technology well. J. P. Geddes's ''Multiple Launch Rocket System to Counter Surge Attack'' (98) examines the U.S. Army's multiple rocket launcher designed for counterbattery fires and suppression of enemy air defenses. Wolfgang Flume and Enrico Po's ''MLRS: An Artillery Rocket System for NATO'' (82) critiques the decision to adopt the American Multiple-Launch Rocket System as a standard NATO rocket system. George T. Norris's ''MRL around the World'' (180) covers the development of multiple rocket launchers in various countries in the 1980s. He discusses ranges, types of warheads, and tactics. G. W. Goodman's ''Jockeying with JTACMS'' (102) is excellent for its examination of the U.S. Army's effort to develop rockets that carry submunitions capable of supporting the deep attack and are fired from the Multiple-Launch Rocket System.

''Multiple Rocket Launcher for the German Armed Forces'' (176) is an excellent discussion of the initial German interest in multiple rocket launchers. For more on German multiple rocket launchers consult D. W. Hoffman's ''A Look at LARS'' (126). R. P. Smith's ''The German MLRS: NATO Commonality of Field Artillery Systems'' (232) is also valuable for its content and reasoning. Ronald T. Pretty's ''South African Valkiri Multiple Artillery Rocket System'' (199) is a short article on South Africa's drive to arm itself with multiple rocket launchers. For a solid discussion of Israeli multiple rocket launchers see Tamir Eshel's ''MLRS in the IDF'' (72). For another critical look at Israeli artillery rockets see Terry J. Gander's ''Israeli Artillery Rockets'' (95). G. A. Halbert's ''Development of Soviet Multiple Rocket Launchers'' (114) provides a useful analysis of the growth of multiple rocket launchers in the Soviet army.

For field artillery technological developments in specific armies between 1919 and 1991 consult Harry G. Bishop's *Field Artillery: The King of Battle* (31). It is the best source available on American field artillery in the 1930s. It also has an excellent bibliography. For additional sources on American field artillery technology consult the U.S. Ordnance Department's *Handbook of Artillery* (262), Leslie E. Babcock's *Elements of Field Artillery* (11), and Arthur R. Wilson's *Field Artillery Manual* (280). The U.S. Army Field Artillery School's *History of the Development of Field Artillery Materiel* (257) has a brief section on field artillery developments in the 1920s and 1930s, especially American. Leslie J. McNair's *Armament of Artillery* (170) is a short, useful examination of the U.S. Army's field artillery immediately following World War I. The U.S. Army Field Artillery School's *Materiel (Elementary)* (259) is critical for understanding the state of American field artillery during the 1930s. The U.S. Army Field Artillery School's *Development of Field Artillery Materiel* (255) is also critical. For coverage of American field artillery during World War II examine William C. Farmer's *Ordnance Field Guide* (77). The U.S. Army Field Artillery School's *Index to Publications Pertaining to Field Artillery Materiel* (258) is an invaluable reference guide.

Larry D. Roberts's ''American Field Artillery: 1930–1939'' (212) provides a critical analysis of American field artillery during a crucial period of transition. John P. Lucas's ''The 105mm Howitzer'' (165) and Janice McKenney's ''More Bang for the Buck in the Interwar Army: The 105mm Howitzer'' (169) are solid articles on the development of a valuable howitzer in the U.S. Army during the 1920s and 1930s. For a sound discussion of the development of the proximity or variable time fuze see Ralph B. Baldwin's *The Deadly Fuze: The Secret Weapons of World War II* (13).

For American rocketry see Nels A. Parson's *Guided Missiles in War and Peace* (190), which explains the origins of the American effort to develop rockets and the theory behind rocketry, and ''Guided Missiles: The Artilleryman's Answer'' (191), which is invaluable for its discussion of early American field artillery rockets. T. B. Hedekin's ''Artillery Rockets'' (122) covers American field artillery rockets during World War II. Holger N. Toftoy's ''Army Missile Development,'' (248) is a critical history of American rocket developments following World War II. One of the best accounts of the U.S. Army's involvement in guided missiles and rockets is ''The Army's Missiles'' (243). This special edition of the *Army Information Digest* covers the origins of the Army's missile and rocket programs following World War II.

A solid article on modernizing the U.S. Army's M109 self-propelled 155mm howitzer in the 1980s is ''Getting HELP and HIP'' (100). It examines the rationale for the U.S. Army's decision to modernize its self-propelled 155mm howitzer. ''BMY's M109 for the 1990s'' (32) is a vital discussion on the proposed modernization of American self-propelled 155mm howitzer. ''Field Artillery Survivability'' (80) is also valuable for the discussion on modernizing the American M109 self-propelled howitzer. In ''Direct Support Field Artillery beyond 1990'' (33), W. W. Breen speculates about the future of American field artillery and the direction in which it needs to go to stay apace of Soviet developments.

Émile Rimailho's *Artillerie de Campagne* (Field Artillery, 208) examines French field artillery materiel during and after World War I. Pierre H. Alvin's *Les Canons de la Victoire* (The Victorious Cannon, 4) is another solid work on French field artillery. For more on French field artillery see Francis A. Leclerc's ''French Artillery'' (158). G. Turbe's ''Far-Reaching Changes in French Artillery'' (249) examines the French effort to modernize and stay abreast of new field artillery technology.

A.P.H.B. Fowle's ''Artillery in the Seventies'' (90) looks at field artillery, especially British, at the beginning of the 1970s. T.D.G. Quayle's ''Fighting the Field Artillery in the British Corps Battle of the 1990s'' (201) provides a solid survey of British tactics, materiel, and organization. ''Bull Launches New Howitzers'' (36) and R. B. Pengelly's ''Artillery Modernization: A British View'' (194) explore the introduction of new field artillery technology during the 1980s.

Literature is also available on German field artillery developments. Written by the former commanding officer of the Peenemünde Rocket Research Institute, Walter Dornberger's *V–2* (63), translated by James Cleugh and Geoffrey Halliday, is an excellent starting point for German rocketry during World War II. Also see A. R. Weyl's ''Development of the V–2'' (276). Basil Collier's *The Battle of the V-Weapons, 1944–1945* (45) is an interesting account of the development of the German V–1 and V–2 weapons, while Peter G. Cooksley's *Flying Bombs: The Story of Hitler's V-Weapons in World War II* (50) re-creates the bombings. David Irving's *The Mare's Nest: The German Secret Weapons Campaign and Allied Countermeasures* (140) and Leslie E. Simon's *German Research in World War II* (227) are other worthwhile books on German rocketry during World War II. Willy Ley's *Rockets: The Future of Travel beyond the Stratosphere* (160) covers German rocketry.

W. Schneider's "Updating the West German Army M109 SP Howitzer" (225) and "German Army Procurement Programs for the 1990s" (224) cover West German modernization activities.

"A Look at Soviet Weapons" (2) discusses Soviet field artillery of the 1950s. Scott R. Gourley's "The Long-Range Threat" (108) is particularly informative on Soviet field artillery technology of the 1970s and 1980s. In "Soviet Artillery: Determined to Dominate the Battlefield" (17) William P. Baxter provides a perceptive analysis of Soviet field artillery technology in the 1970s and early 1980s. A useful study on Soviet field artillery materiel from the 1980s onward is Christopher Bellamy's "Destruction by Fire: Soviet Artillery in the 1980s and Beyond" (19). The same author's "Firepower Superiority" (20) is also a solid discussion of Soviet field artillery. C. N. Donnelly's "Modern Soviet Artillery" (57) is another excellent article on Soviet field artillery.

For Soviet self-propelled artillery of the 1970s and 1980s see Steven J. Zaloga's "The Mechanized Threat" (283). For more coverage of Soviet field artillery consult D. Eshel's "Soviet Self-Propelled Artillery" (71). Less analytical is Andrew W. Hull's "Evolution of Soviet Self-Propelled Artillery" (137).

I. A. Slukhai's *Russian Rocketry: A Historical Survey* (228), translated by E. Vilam, furnishes an overview of Russian rocket history. For a brief examination of Soviet rockets and missiles during the 1950s consult "The Soviet Union's Missile Artillery" (246). William P. Baxter's "Soviet Economy of Massed Fires" (18) studies multiple rocket launcher technology in the Soviet army from the 1960s to 1980s. Another fine study on Soviet multiple rocket launcher technology is G. A. Halbert's "Development of Soviet Multiple Rocket Launchers" (114).

In "Artillery in the Middle East" (103) John Gordon provides a useful examination of the guns of each Middle Eastern country in the 1980s, while Scott R. Gourley's "The Asian Artillery Threat" (107) is critical for the state of field artillery in Asia in the 1980s. Christopher F. Foss's "Recent Developments in Chinese Artillery" (89) is vital.

TACTICS, DOCTRINE, AND ORGANIZATION

For a brief examination of field artillery tactics in the twentieth century consult J. C. Antoniotti's "Field Artillery: Its Role in Ground Warfare before 1990" (5). École Supérieure de Guerre's *Cours de Tactique d'Artillerie* (A Course in Artillery Tactics, 69) discusses French field artillery tactics on the basis of World War I experiences. In *Artillery Tactics: 1939–1945* (28) Shelford Bidwell perceptively analyzes field artillery tactics as they evolved during World War II.

Books available on field artillery tactics since World War II are scarce. For the most part, journal articles are the best source for the subject. G. C. Reinhardt's "Notes on the Tactical Employment of Atomic Weapons" (206) is invaluable because the author is one of the first to consider tactical nuclear warfare. G. C. Reinhardt and W. R. Kintner's *Atomic Weapons in Land Combat* (207) is a pioneering work on atomic field artillery tactics in the U.S. Army. Another thoughtful examination of atomic artillery is furnished in Ferdinand O. Miksche's *Atomic Weapons and Armies* (171). M. St. J. Oswald's "Field Artillery in Atomic Warfare" (185) is essential for the initial concepts of employing tactical nuclear field artillery on the battlefield.

For an examination of field artillery tactics during the Arab-Israeli War of 1967 consult D. D. Campbell's "The Gunners in the Arab-Israeli War of 1967" (40). T. L. Morony's "Artillery Support in the Yom Kippur War" (175) should also be consulted for that particular war. An interesting perspective on the impact of technology on the battlefield is Julia Ackerman's "Bold New Doctrine, Weapons for the High-Tech Battlefield" (1). The article covers new doctrine and tactics for field artillery of the 1980s. For an invaluable discussion on the impact of precision-guided artillery munitions on the battlefield and tactics see Ferdinand O. Miksche's "PGMs Are Changing the Combat Picture" (172).

For German field artillery tactics, doctrine, and tactics examine Friedrich Feeser's *Artillerie im Feldkrieg* (Artillery in War, 78). The Kriegsministerium's *Artillery Command* (155), translated by Joseph Dasher, is another basic source for German field artillery tactics on the eve of World War II. The Kriegsministerium's *Ausbildung einer Batterie und Ausbildungsvorschrift für die Artillerie* (Training a Battery and Training Regulations for the Artillery, 156) is also valuable for German tactics as well as training.

Raymond L. Garthoff's *Soviet Military Doctrine* (96) is a starting place for Soviet doctrine and complements works on Soviet artillery tactics and doctrine. The U.S. Army Command and General Staff College's *Organization and Employment of a Division, Soviet (Russian) Army* (251) furnishes a combined arms perspective, as does the U.S. War Department Military Intelligence Service's *The Organization and Tactical Employment of Soviet Ground Forces (July 1944)* (272). V. D. Grendall's *Artillery in Principal Roles of Combat* (111) focuses on Soviet field artillery tactics. Although the U.S. Army translated the title from its original Russian, the text is still in Russian. The U.S.S.R.'s *Course in Artillery, Book One* (263) and *Course in Artillery, Book Four* (264) are sound sources, as is the U.S.S.R. Ministry of Armed Force's *Course in Artillery* (265). These three books on Soviet artillery tactics were translated by the U.S. Army.

Works on recent Soviet field artillery tactics, doctrine, and organization are also accessible. C. N. Donnelly's "Tactical Problems Facing the Soviet Army: Part One" (58) and "Tactical Problems Facing the Soviet Army: Part Two" (59) are worth examining for their insights into Soviet tactics, including field artillery. David C. Isby's "Afghanistan: The Threat of Soviet Artillery," (142) furnishes an outline of Soviet artillery technology and tactics. The same author's "Soviet Self-Propelled Artillery: Capabilities and Concepts" (141) is another vital work on Soviet doctrine. G. Jacobs's "Soviet War Concept: Modernization's Impact" (144) is another solid examination of the impact of new field artillery technology, as well as technology in general, upon tactics and doctrine. Another worthwhile article on Soviet field artillery tactics is C. N. Donnelly's "The Development of the Soviet Concept of Echeloning" (60).

For field artillery tactics in the British army consult A. A. Goschen's "Artillery Tactics" (106). Also see Robert Scales's "Firepower in the Falkland Campaign" (223). D. S. Gay's "The Survival of Our Guns: A Concept of Future Development" (97) is a useful discussion of British tactics and doctrine, as is D.W.L. Robinson's "Concept of Employment of Artillery in the 1980s" (213). Robinson's article is particularly worthwhile because the author envisions what field artillery doctrine would be like in the 1980s.

The U.S. Cavalry School's *Other Arms: Field Artillery with Cavalry* (261) is essential for American field artillery tactics during the early 1930s. The U.S. Army Command and General Staff College's *Tactics and Techniques: Artillery* (252), *Tactics and Techniques: Artillery* (253), and *Division Artillery* (250) are invaluable primary sources. All of the books cover tactics, organization, and equipment. The U.S. Army Field Artillery School's *Field Artillery Fundamentals* (256) and *Tactical Employment of Field Artillery* (260) provide solid foundations for understanding American field artillery tactics and organization during World War II.

For a broad coverage of U.S. Army doctrine and tactics see John L. Romjue's *From Active Defense to AirLand Battle: The Development of Army Doctrine, 1973–1982* (216). It is the standard source for the development of AirLand Battle during the late 1970s and early 1980s. S. G. Starner's "Deep Attack: We Can Do It Now" (235) is important for its view that AirLand Battle is a reality. "Implementing the AirLand Battle" (138) is an invaluable article on fighting AirLand Battle in Europe. Bill Rittenhouse's "It's Deeper Than You Think" (210) examines AirLand Battle from a field artillery perspective. For a solid discussion on the challenges of defeating Soviet artillery see Eddy Smith's "Defeating Soviet Artillery" (230), which has a fine bibliography. David L. Ingle's "The Role of Corps Artillery in the AirLand Battle Concept" (139) is worth reading for its discussion about the change in the role of corps artillery in the American field artillery and its fine bibliography on a much-neglected topic.

Serialized extracts of David E. Ott's *Field Artillery: 1954–1973* (186) can be found in a seven-part "U.S. Field Artillery in Vietnam" (187) and furnish a solid analysis of American field artillery tactics, organization, and materiel. L. J. Picou's "Airmobile Artillery in Combat" (195) provides a valuable discussion of aerial rocket artillery in the first years of the Vietnam War. Robert H. Scales's "Firepower and Maneuver in the Second Indochina War" (222) offers insights into field artillery employment in small wars.

BIBLIOGRAPHY

1. Ackerman, Julia. "Bold New Doctrine, Weapons for the High-Tech Battlefield." *Defense and Foreign Affairs*, June 1985, 16–19, 22–23.

2. "A Look at Soviet Weapons." *Army Information Digest*, August 1957, 2–14.
3. "Allied Artillery Missiles." *Artillery Trends*, February 1960, 16–20.
4. Alvin, Pierre H. *Les Canons de la Victoire*. Paris: Charles-Lavauzelle, 1923.
5. Antoniotti, J. C. "Field Artillery: Its Role in Ground Warfare before 1990." *NATO's 16 Nations*, August-September 1983, 69–72.
6. ———. "Precision Guided Munitions." *International Defense Review*, September 1986, 7–11.
7. Appleman, Roy E. *South to the Naktong, North to the Yalu*. Washington, DC: Office of the Chief of Military History, U.S. Army, 1961.
8. Arnold, James R. *The Illustrated History of Artillery: The Vietnam War*. New York: Bantam Books, 1987.
9. "Artillery Command and Control." *International Defense Review*, January 1985, 39–61.
10. "Artillery Rocket Systems: A Unique Market." *Military Technology*, September 1987, 22–31.
11. Babcock, Leslie E. *Elements of Field Artillery*. Princeton, NJ: Princeton University Press, 1925.
12. Bailey, Jonathan B.A. *Field Artillery and Firepower*. Oxford, England: The Military Press, 1989.
13. Baldwin, Ralph B. *The Deadly Fuze: The Secret Weapons of World War II*. San Rafael, CA: Presidio Press, 1980.
14. Barnes, Leslie W. *Canada's Guns: An Illustrated History of Artillery*. Ottawa: National Museum of Man, National Museum of Canada, 1979.
15. Batchelor, John, and Ian V. Hogg. *Artillery*. New York: Charles Scribner's Sons, 1972.
16. "BATES: A Layman's Guide." *International Defense Review*, February 1978, 104.
17. Baxter, William P. "Soviet Artillery: Determined to Dominate the Battlefield." *Army*, April 1982, 57–59.
18. ———. "Soviet Economy of Massed Fires." *Army*, June 1982, 61–63.
19. Bellamy, Christopher. "Destruction by Fire: Soviet Artillery in the 1980s and Beyond." *Field Artillery Journal*, September-October 1985, 45–50.
20. ———. "Firepower Superiority." *Jane's Defence Weekly*, 12 January 1985, 59–66.
21. ———. *Red God of War: Soviet Artillery and Rocket Forces*. Washington, DC: Brassey's Defence Publishers, 1986.
22. ———. "Soviet Artillery and Rocket Forces: 1940–1980." *Jane's Defence Weekly* 4, 3 (1983): 269–80.
23. Berman, Larry. *Planning a Tragedy: The Americanization of the War in Vietnam*. New York: W. W. Norton, 1982.
24. Bidwell, R.G.S. "The Development of British Field Artillery Tactics, 1920–1939." *Journal of the Royal Artillery*, March 1967, 13–24.
25. ———. "The Development of British Field Artillery Tactics, 1940–1942." *Journal of the Royal Artillery*, September 1967, 1–13.
26. ———. "The Development of British Field Artillery Tactics, 1940–1943." *Journal of the Royal Artillery*, March 1968, 1–13.
27. Bidwell, Shelford. *Gunners at War: A Tactical Study of the Royal Artillery in the Twentieth Century*. London: Arms and Armour Press, 1970.

28. ———. *Artillery Tactics: 1939–1945*. London: Almark Publishing Company, 1976.
29. ———, ed. *Brassey's Artillery of the World*. New York: Bonanza Press, 1979.
30. Bidwell, Shelford, and Dominick Graham. *Firepower: British Army Weapons and Theories of War: 1904–1945*. London: Allen and Unwin, 1982.
31. Bishop, Harry G. *Field Artillery: The King of Battle*. Boston: Houghton Mifflin, 1935.
32. "BMY's M109 for the 1990s." *Defense Attaché*, 6/1984, 21–22.
33. Breen, W. W. "Direct Support Field Artillery beyond 1990." *Field Artillery Journal*, July-August 1980, 33–39.
34. Brenner, Eliot, and William Harwood. *Desert Storm: The Weapons of War*. New York: Crown Publishers, 1991.
35. Brookes, Kenneth. *Battle Thunder: The Story of Britain's Artillery*. Berkshire, England: Osprey Publishing Ltd., 1979.
36. "Bull Launches New Howitzers." *International Defense Review*, February 1987, 169–70.
37. Burchard, John E., ed. *Rockets, Guns, and Targets*. Boston: Little, Brown, 1948.
38. Bursell, William R. "American Sound Ranging in Four Wars." *Field Artillery Journal*, November-December 1981, 53–55.
39. Campbell, Christopher. *Nuclear Weapons Fact Book*. Novato, CA: Presidio Press, 1984.
40. Campbell, D. D. "The Gunners in the Arab-Israeli War of 1967." *Journal of the Royal Artillery*, September 1968, 132–37.
41. Carver, Michael. *War since 1945*. New York: G. P. Putnam, 1980.
42. Chamberlain, Peter. *Heavy Artillery: World War II Fact File*. New York: Arco Publishing, 1975.
43. ———. *Light and Medium Field Artillery: World War II Fact File*. New York: Arco Publishing, 1975.
44. Cole, A. G. "German Artillery Concentrations in World War II." *Royal Artillery Journal*, 1948, 196–99.
45. Collier, Basil. *The Battle of the V-Weapons, 1944–1945*. New York: William Morrow, 1965.
46. Comparato, Frank E. *Age of Great Guns: Cannon Kings and Cannoneers Who Forged the Firepower of Artillery*. Harrisburg, PA: The Stackpole Company, 1965.
47. "Computers and Artillery: From FACE to BATES." *Miltronics*, December 1984, 187–92.
48. "Concept for Self-propelled Howitzer." *Jane's Defence Weekly*, 31 January 1987, 162.
49. "Considerations on the Modern Tube Artillery." *Military Technology*, April 1983, 52–62.
50. Cooksley, Peter G. *Flying Bombs: The Story of Hitler's V-Weapons in World War II*. New York: Charles Scribner's Sons, 1979.
51. Cordesman, Anthony H., and Abraham R. Wagner. *The Lessons of Modern War*. 3 vols. Boulder, CO: Westview Press, 1990.
52. Crane, J. A. "Field Artillery Groups." *Field Artillery Journal*, October 1945, 579–82.
53. Craven, R.C.F. "A70 for the 90s." *Field Artillery Journal*, May-June 1983, 26–31.

54. Crutchley, Michael J., and John H. Milam. "The Modernisation of NATO's Tube Artillery." *Military Technology*, October 1985, 28–39.
55. Dastrup, Boyd L. *King of Battle: A Branch History of the U.S. Army's Field Artillery*. Fort Monroe, VA: Office of the Command Historian, U.S. Army Training and Doctrine Command, 1992.
56. Dodd, Norman. "The Royal Artillery of the British Army." *Field Artillery Journal*, January-February 1974, 6–14.
57. Donnelly, C. N. "Modern Soviet Artillery." *NATO's 15 Nations*, June-July 1979, 48–54.
58. ———. "Tactical Problems Facing the Soviet Army: Part One." *Military Review*, June 1979, 18–26.
59. ———. "Tactical Problems Facing the Soviet Army: Part Two." *Military Review*, July 1979, 60–68.
60. ———. "The Development of the Soviet Concept of Echeloning." *NATO Review*, December 1984, 9–17.
61. Donnelly, C. H. "The Artillery Aerial OP." *Military Review*, April 1943, 36–37.
62. Donnelly, Ralph W. *An Annotated Bibliography of the United States Marine Corps' Artillery*. Washington, DC: U.S. Marine Corps, 1970.
63. Dornberger, Walter. *V–2*. Translated by James Cleugh and Geoffrey Halliday. New York: Viking Press, 1954.
64. Doughty, Robert A. *The Evolution of U.S. Army Tactical Doctrine, 1946–1976*. Fort Leavenworth, KS: Combat Studies Institute, U.S. Army Command and General Staff College, 1979.
65. Downey, Fairfax. *Cannonade: Great Artillery Actions of History, the Famous Cannons, and the Master Gunners*. Garden City, NY: Doubleday, 1966.
66. ———. *Sound of the Guns: The Story of American Artillery*. New York: David McKay, 1955.
67. Drake-Brockman, V. "Modern Developments in Foreign Artilleries of the Field Army." *Journal of the Royal Artillery*, 1924–1925, 53–64.
68. Dunnigan, James F., and Austin Bay. *From Shield to Storm: High Tech Weapons, Military Strategy, and Coalition Warfare in the Persian Gulf*. New York: William Morrow, 1991.
69. École Supérieure de Guerre. *Cours de Tactique d'Artillerie*. Paris: École Supérieure de Guerre, 1925.
70. Epstein, Laurence B. "Army Organic Light Aviation: The Founding Fathers." *U.S. Army Aviation Digest*, June 1977, 2–17.
71. Eshel, D. "Soviet Self-Propelled Artillery." *Defense Update International* 49 (1984); 37–41.
72. Eshel, Tamir. "MLRS in the IDF." *Defense Update International*, 1984, 58–63.
73. Esposito, V. J. *A Concise History of World War II*. New York: Praeger, 1964.
74. Evans, I. D. "Artillery on the Battlefield: The Future of Towed Guns." *NATO's 15 Nations*, August-September 1981, 76–81.
75. Evans, N. F. "The Impact of BATES on the Royal Artillery." *Journal of the Royal Artillery*, September 1987, 131–35.
76. Falls, Bernard. *Street without Joy*. 4th ed. Harrisburg, PA: The Stackpole Company, 1964.
77. Farmer, William C. *Ordnance Field Guide*. Harrisburg, PA: Military Service Publishing Company, 1944.

78. Feeser, Friedrich. *Artillerie im Feldkrieg*. Berlin: E. S. Mittler und Sohn, 1930.
79. Fehrenbach, T. R. *This Kind of War: A Study in Unpreparedness*. New York: Macmillan, 1963.
80. "Field Artillery Survivability." *Field Artillery Journal*, May-June 1980, 8–21.
81. Flume, Wolfgang. "Advanced Artillery Munitions: A German View." *Military Technology*, December 1985, 55–61.
82. Flume, Wolfgang, and Enrico Po. "MLRS: An Artillery Rocket System for NATO." *Military Technology*, February 1985, 15–25.
83. Ford, William W. "Grasshoppers." *U.S. Army Aviation Digest*, June 1982, 2–10.
84. ———. *Wagon Soldiers*. West Redding, CT: Excelsior Printing Company, 1980.
85. Forward, I. K. "The Case against Copperhead." *Journal of the Royal Artillery*, March 1984, 29–31.
86. Foss, Christopher F. *Artillery of the World*. New York: Charles Scribner's Sons, 1974.
87. ———. *Jane's Pocket Book of Towed Artillery*. New York: Collier Books, 1979.
88. ———. "Munitions Development." *Jane's Defence Weekly*, 13 December 1986, 1389–92.
89. ———. "Recent Developments in Chinese Artillery." *Jane's Defence Weekly*, 27 December 1986, 1484–86.
90. Fowles, A.P.H.B. "Artillery in the Seventies." *Journal of the Royal Artillery*, September 1973, 81–89.
91. Friedman, Norman. *Desert Victory: The War for Kuwait*. Annapolis, MD: Naval Institute Press, 1991.
92. Furlong, R.D.M. "Self-Propelled 155mm Artillery for the 90s." *International Defense Review*, September 1983, 1253–64.
93. Gahagan, Larry D. "Development and Use of Field Artillery Proximity Fuzes in World War II." *Field Artillery Journal*, March-April 1980, 15–18.
94. Gander, Terry J. *Artillery*. Minneapolis: Lerner Publications, 1986.
95. ———. "Israeli Artillery Rockets." *Jane's Defence Weekly*, July 1984, 1109–10.
96. Garthoff, Raymond L. *Soviet Military Doctrine*. Glencoe, IL: The Free Press, 1953.
97. Gay, D. S. "The Survival of Our Guns: A Concept of Future Development." *Journal of the Royal Artillery*, March 1986, 5–10.
98. Geddes, J. P. "Multiple Launch Rocket System to Counter Surge Attack." *International Defense Review*, May 1980, 728–32.
99. "German RPV Programs." *International Defense Review*, November 1985, 1781–82.
100. "Getting HELP and HIP." *Field Artillery Journal*, September-October 1985, 20–21.
101. Gleckler, Jim. *Redleg*. Miami, OK: Northeastern Oklahoma A & M College, 1985.
101A. Goodman, A. Sue, ed. *Persian Gulf War, 1990–1991*. Maxwell Air Force Base, AL: Special Bibliography no. 297, Air University Library, 1991.
102. Goodman, G. W. "Jockeying with JTACMS." *Field Artillery Journal*, January-February 1986, 26–27.
103. Gordon, John. "Artillery in the Middle East." *Field Artillery Journal*, January-February 1987, 24–28.
104. ———. "The Evolution of Soviet Fire Support, 1940–1988." *Field Artillery*, June 1988, 18–21.

105. Gorin, Lewis J., Jr. *The Cannon's Mouth: The Role of U.S. Artillery during World War II*. New York: Carlton Press, 1973.
106. Goschen, A. A. "Artillery Tactics." *Journal of the Royal Artillery*, July 1925, 254–60.
107. Gourley, Scott R. "The Asian Artillery Threat." *Field Artillery Journal*, January-February 1987, 13–19.
108. ———. "The Long-Range Threat." *Field Artillery Journal*. January-February 1987, 14–17.
109. Graham, C.A.L. *The History of the Indian Mountain Artillery*. Aldershot, England: Gale and Polden Ltd., 1957.
110. Gray, A. J. "Automatic Data Processing in the Field Artillery in the 1980s." *Journal of the Royal Artillery*, March 1978, 46–52.
111. Grendall, V. D. *Artillery in Principal Roles of Combat*. [English title. Text in Russian.] Moscow: Government Printing Office, People's Commissariat of Defense of U.S.S.R., 1940.
112. Griffith, Paddy. *The Ultimate Weaponry*. New York: St. Martin's Press, 1991.
112A. Gudmundsson, Bruce I. *On Artillery*, Westport, CT: Greenwood Press, 1993.
113. Gugeler, Russell A. *Combat Actions in Korea: Infantry, Artillery, Armor*. Washington, DC: Combat Forces Press, 1954.
114. Halbert, G. A. "Development of Soviet Multiple Rocket Launchers." *Field Artillery Journal*, August 1987, 39–41.
115. Hall, James F. "Precision Guided Artillery: First and Second Generation Projectiles." *Field Artillery Journal*, May-June 1981, 9–12.
116. Hammersen, Frederick P. A. "A History of Russian and Soviet Artillery, 1914–1941." Master's thesis, University of Virginia, 1977.
117. Harrell, I. J. "Development and Employment of Rockets." *Military Review*, November 1945, 24–28.
118. Hart, C. E. "Artillery with an American Army in Europe." *Field Artillery Journal*, September 1945, 23–27.
119. Hart, D. M., and E. M. Gormley, "The Evolution of Soviet Interest in Atomic Artillery." *Journal of the Royal United Services Institute for Defence Studies*, June 1983, 25–34.
120. Harvey, D. L. "A Troubled Nest for Aquila." *NATO's 16 Nations*, April 1986, 74–76.
121. Hayes, Thomas. *Elements of Ordnance*. New York: John Wiley and Sons, 1938.
122. Hedekin, T. B. "Artillery Rockets." *Field Artillery Journal*, October 1946, 564–73.
123. Herring, George C. *America's Longest War: The United States and Vietnam, 1950–1975*. New York: John Wiley and Sons, 1979.
124. Herzog, Chaim. *The Arab-Israeli Wars: War and Peace in the Middle East from the War of Independence through Lebanon*. New York: Vintage Books, 1984.
125. Higham, Robin, ed. *Official Histories: Essays and Bibliographies from Around the World*. Manhattan, KS: Kansas State University Library, 1970.
126. Hoffman, D. W. "A Look at LARS." *Field Artillery Journal*, September-October 1985, 13.
127. ———. "Developments in Precision Guided Munitions." *NATO's 16 Nations*, April-May 1985, 48–55.

128. Hoffman, K. "An Analysis of Soviet Artillery Development." *International Defense Review*. December 1977, 5–10.
129. Hogg, Ian V. *Artillery 2000*. London: Arms and Armour Press, 1990.
130. ———. *Artillery in Color: 1920–1963*. New York: Arco Publishing, 1980.
131. ———. *British and American Artillery of World War II*. London: Arms and Armour Press, 1978.
132. ———. *German Artillery of World War II*. London: Arms and Armour Press, 1975.
133. ———. *The Guns: 1939–1945*. New York: Ballantine Books, 1970.
134. ———. *The Illustrated Encyclopedia of Artillery*. London: Guild Publishing, 1987.
135. Hogg, O.F.G. *Artillery: Its Origin, Heyday, and Decline*. Hamden, CT: Archon Books, 1970.
136. Holthus, Michael D., and Steven M. Chandler. "Myths and Lessons of Iraqi Artillery." *Field Artillery*, October 1991, 7–9.
137. Hull, Andrew W. "Evolution of Soviet Self-propelled Artillery." *Field Artillery Journal*, March-April 1978, 8–12, 33.
138. "Implementing the AirLand Battle." *Field Artillery Journal*, September-October 1981, 20–27.
139. Ingle, David L. "The Role of Corps Artillery in the AirLand Battle Concept." Master's thesis, U.S. Army Command and General Staff College, 1982.
140. Irving, David. *The Mare's Nest: The German Secret Weapons Campaign and Allied Countermeasures*. London: W. Kimber, 1964.
141. Isby, David C. "Soviet Self-Propelled Artillery: Capabilities and Concepts." *Jane's Defence Review*, 1981, 200–204.
142. ———. "Afghanistan: The Threat of Soviet Artillery." *Field Artillery Journal*, January-February 1987, 8–12.
143. Jackson, T. J. "Precision Guided Munitions for the U.S. Army." *Military Technology*, December 1985, 44–54.
144. Jacobs, G. "Soviet War Concept: Modernization's Impact." *Asian Defense Journal*, August 1984, 24–39.
145. *Jane's Armour and Artillery*. London: Jane's Publications, 1979.
146. *Jane's Armour and Artillery*. London: Jane's Publications, 1980–1989.
147. *Jane's Weapons Systems*. London: Jane's Publications, 1980–1989.
148. Jobe, Joseph, ed. *Guns: An Illustrated History of Artillery*. Greenwich, CT: New York Graphic Society, 1971.
149. Johnson, Curt. *Artillery: The Big Guns Go to War*. London: Octopus Books, 1975.
150. Kamarck, A. "Rebirth of the War Rocket." *Field Artillery Journal*, July 1943, 506–9.
151. Karnow, Stanley. *Vietnam: A History*. New York: Viking Press, 1983.
152. Kelly, James P., ed. *Field Artillery Materiel*. Columbia, MO: University of Missouri Press, 1920.
153. Kimche, David, and Dan Bawly. *The Six-Day War: Prologue and Aftermath*. New York: Stein and Day, 1971.
154. Kosar, Franz. *A Pocket History of Artillery: Light Field Guns*. London: Ian Allen, 1971.
155. Kriegsministerium. *Artillery Command*. Translated by Joseph Dasher. Washington, DC: Office of the Chief of Field Artillery, 1940.
156. ———. *Ausbildung einer Batterie und Ausbildungsvorschrift für die Artillerie*. 3 vols. Berlin: E. S. Mittler and Sohn, 1925–1928.

157. Leckie, Robert. *Conflict: The History of the Korean War*. New York: G. P. Putnam, 1962.
158. Leclerc, Francis A. "French Artillery." *Field Artillery Journal*, July-August 1973, 24–39.
159. Lewy, Guenther. *America in Vietnam*. New York: Oxford University Press, 1978.
160. Ley, Willy. *Rockets: The Future of Travel beyond the Stratosphere*. New York: Viking Press, 1944.
161. Liddell Hart, B. H. *History of the Second World War*. New York: G. P. Putnam, 1971.
162. Lingamfelter, L. Scott. "In the Wake of the Storm: Improving Field Artillery after Operation Desert Storm." *Field Artillery*, August 1991, 27–29.
163. Lippincott, W. R. "RPV: Tomorrow's Armchair Giant Killer." *Army*, May 1976, 27–32.
164. Lorber, A. K. "Active Defense against Precision Guided Munitions for the Future Battlefield." *Military Technology*, May 1985, 77–82.
165. Lucas, John P. "The 105mm Howitzer." *Field Artillery Journal*, February 1941, 66–69.
166. McClean, Donald B. *Illustrated Arsenal of the Third Reich*. Wickenburg, AZ: Normount Technical Publications, 1973.
167. McFarland, Earl. *Textbook of Ordnance and Gunnery*. New York: John Wiley and Sons, 1929.
168. McKenney, Janice A. *Field Artillery*. Washington, DC: U.S. Army Center of Military History, 1986.
169. ———. "More Bang for the Buck in the Interwar Army: The 105mm Howitzer." *Military Affairs*, April 1978, 80–86.
170. McNair, Leslie J. *Armament of Artillery*. Washington, DC, n.p., 1919.
170A. Menos, Dennis. *Arms over Diplomacy: Reflections on the Persian Gulf War*. Westport, CT: Praeger, 1992.
171. Miksche, Ferdinand Otto. *Atomic Weapons and Armies*. New York: Praeger, 1955.
172. ———. "PGMs Are Changing the Combat Picture." *Military Review*, July 1978, 10–18.
173. Millet, Allan R., and Williamson Murray. *Military Effectiveness*. 3 vols. Boston: Allen and Unwin, 1987.
174. Mills, Phillip J. "RPVs over the Bekaa Valley." *Army*, June 1983, 49–51.
175. Morony, T. L. "Artillery Support in the Yom Kippur War." *Journal of the Royal Artillery*, March 1974, 11–14.
176. "Multiple Rocket Launcher for the German Armed Forces." *International Defense Review*, April 1968, 60–65.
177. "New Generation Weapons and the Doctrine of Tactics." *Jane's Defence Weekly*, 14 June 1986, 1102–6.
178. Nicholson, G.W.L. *The Gunners of Canada: The History of the Royal Regiment of Canadian Artillery*. Toronto: McClelland and Stewart, 1967.
179. Niemzig, Ngo. "The SP–70–1." *NATO's 15 Nations*, October-November 1981, 82–90.
180. Norris, George T. "MRL around the World." *Field Artillery Journal*, August 1987, 37–38.
181. Ogorkiewicz, R. M. "The Evolution of Self-propelled Guns." *Military Review*, October 1951, 102–9.

182. O'Hagan, M. L. "BATES: A Layman's Guide." *Journal of the Royal Artillery*, September 1980, 122–25.

183. ———. *Missiles and the Revolution in Warfare*. Cambridge, MA: Harvard University Press, 1962.

184. O'Malley, T. J. "Self-Propelled Artillery for the Late 1980s." *Defence Today*, April 1985, 96–106.

185. Oswald, M. St. J. "Field Artillery in Atomic Warfare." *Journal of the Royal Artillery*, January 1958, 46–52.

186. Ott, David E. *Field Artillery: 1954–1973*. Washington, DC: Department of the Army, 1975.

187. ———. "U.S. Field Artillery in Vietnam." *Field Artillery Journal*, January-February 1975, 24–29; March-April 1975, 39–45; May-June 1975, 15–22; September-October 1975, 46–51; November-December 1975, 29–34; January-February 1976, 27–31; March-April 1976, 44–48; May-June 1976, 53–57; July-August 1976, 37–43; September-October 1976, 10–17; November-December 1976, 13–18; January-February 1977, 42–47; March-April 1977, 29–33.

188. Palmer, David R. *Summons of the Trumpet*. San Francisco: Presidio Press, 1978.

189. Parham, H. J. *Unarmed into Battle: The Story of the Air Observation Post*. Winchester, England: Wykeham Press, 1956.

190. Parson, Nels A. *Guided Missiles in War and Peace*. Cambridge, MA: Harvard University Press, 1956.

191. ———. "Guided Missiles: The Artilleryman's Answer." *Field Artillery Journal*, May-June 1949, 112–19.

192. Pemberton, A. L. *The Development of Artillery Tactics and Equipment*. London: The War Office, 1950.

193. Pendray, George E. *The Coming of the Age of Rocket Power*. Rev. ed. New York: Harper and Brothers, 1947.

194. Pengelly, R. B. "Artillery Modernization: A British View." *International Defense Review*, February 1987, 179–82.

195. Picou, L. J. "Airmobile Artillery in Combat." *Artillery Trends*, August 1967, 12–20.

196. Pillsbury, Hobart B. "Artillery: New Developments." *Journal of Defense and Diplomacy*, April 1984, 49–53.

197. Pope, Dudley. *Guns: An Illustrated History of Artillery*. New York: Delacorte Press, 1965.

198. Prehar, Bohdan. "Artillery Scatterable Mines." *Field Artillery Journal*, September-October 1979, 8–11.

199. Pretty, Ronald T. "South African Valkiri Multiple Artillery Rocket System." *Jane's Defence Review*, 1983, 645–49.

200. Prochko, I. S. *Artillery: God of War*. Moscow: Soviet News. 1945.

201. Quayle, T.D.G. "Fighting the Field Artillery in the British Corps Battle of the 1990s." *Field Artillery*, February 1990, 15–21.

202. Ratliff, Frank G. "The Field Artillery Fire Direction Center: Its Past, Present, and Future." *Field Artillery Journal*, May-June 1950, 116–19, 127.

203. Reed, John. "Beyond Copperhead: What Prospects for SADARM and STAFF?" *Defence*, January 1985, 35.

204. Rees, David. *Korea: The Limited War*. New York: St. Martin's Press, 1964.

205. Reeves, J. R. ''Artillery in the Ardennes.'' *Field Artillery Journal*, March 1946, 138–42, 173–84.
206. Reinhardt, G. C. ''Notes on the Tactical Employment of Atomic Weapons.'' *Military Review*, September 1952, 28–37.
207. Reinhardt, G. C. and W. R. Kintner. *Atomic Weapons in Land Combat*. Harrisburg, PA: Military Service Publishing Company, 1954.
208. Rimailho, Émile. *Artillerie de Campagne*. Paris: Gauthier-Villars, 1924.
209. Ringer, Matthew J., and Martell D. Fritz. ''FADAC Computations versus Manual Computations.'' *Artillery Trends*, April 1969, 40–42.
210. Rittenhouse, Bill. ''It's Deeper Than You Think.'' *Field Artillery Journal*, January-February 1986, 17–20.
211. ———. ''MLRS: Smart Munitions.'' *Field Artillery*, August 1987, 46–48.
212. Roberts, Larry D. ''American Field Artillery: 1930–1939.'' Master's thesis, Oklahoma State University, 1977.
213. Robinson, D.W.L. ''Concept of Employment of Artillery in the 1980s.'' *Journal of the Royal Artillery*, September 1977, 93–96.
214. ———. ''Why Rockets?'' *Journal of the Royal Artillery*, September 1978, 115–18.
215. Rogers, H.C.B. *A History of Artillery*. Secaucus, NJ: The Citadel Press, 1975.
216. Romjue, John L. *From Active Defense to AirLand Battle: The Development of Army Doctrine, 1973–1982*. Fort Monroe, VA: U.S. Army Training and Doctrine Command Historical Office, 1984.
217. Rowan-Robinson, H. *Artillery: Today and Tomorrow*. London: Clowes and Sons Ltd., 1928.
218. Ryan, J. W. *Guns, Mortars, and Rockets*. New York: Brassey's Publications Ltd., 1982.
219. ''SADARM Scores Twice.'' *Defense Electronics*, June 1985, 47–50.
220. Sawicki, James A. *Field Artillery Battalions of the U.S. Army*. Drumfries, VA: Centaur Publications, 1977.
221. Scales, Robert F. *Firepower in Limited War*. Washington, DC: National Defense University Press, 1990.
222. ———. ''Firepower and Maneuver in the Second Indochina War.'' *Field Artillery Journal*, September-October 1986, 47–53.
223. ———. ''Firepower in the Falkland Campaign.'' *Field Artillery Journal*, May-June 1986, 14–21.
224. Schneider, W. ''Germany Army Procurement Programs for the 1990s.'' *International Defense Review*, December 1986, 1759–61.
225. ———. ''Updating the West German Army M109 SP Howitzer.'' *International Defense Review*, September 1985, 2024.
226. Seidenmann, P., and D. J. Spanovich. ''SADARM.'' *National Defense*, October 1985, 26–30.
227. Simon, Leslie E. *German Research in World War II*. New York: John Wiley and Sons, 1947.
228. Slukhai, I. A. *Russian Rocketry: A Historical Survey*. Translated by E. Vilam. Jerusalem: Israel Program for Scientific Translations, 1968.
229. Smirnov, V. ''Soviet Artillery in World War II.'' *Military Review*, April 1945, pp. 103–7.

230. Smith, Eddy. "Defeating Soviet Artillery." Master's thesis, U.S. Army Command and General Staff College, 1979.

231. Smith, Rupert. "The Gulf War: The Land Battle." *RUSI Journal*, February 1992, 1–5.

232. Smith, R. P. "The German MLRS: NATO Commonality of Field Artillery Systems." *Forward Observer*, July-August 1987, 8–9.

233. Snyder, L. L. *The War: A Concise History, 1939–1945*. New York: J. Messner, 1960.

234. Stambler, Irwin. "U.S. Remotely Piloted Vehicle Program." *International Defense Review*, February 1974, 179–86.

235. Starner, S. G. "Deep Attack: We Can Do It Now." *Field Artillery Journal*, May-June 1986, 11–13.

236. Stevens, Phillips H. *Artillery through the Ages*. Secaucus, NJ: The Citadel Press, 1965.

237. Stokesbury, James L. *A Short History of World War II*. New York: William Morrow, 1980.

238. Stone, Walter. "Current Trends in Artillery." *Defence*, November 1985, 510–14.

239. Sundaram, Gowri S. "The U.S. Tube-Launched Smart Anti-Tank Munitions." *International Defense Review*, August 1980, 1234–36.

240. Sweetman, Bill. "Unmanned Air Vehicles Make a Comeback." *International Defense Review*, November 1985, 1771–77.

241. "TACFIRE: An Automated Artillery Fire Direction System." *International Defense Review*, February 1971, 72–76.

242. Tauson, G. "Artillery Development in World War II." *Military Review*, March 1948, 91–97.

243. "The Army's Missiles." *Army Information Digest*, December 1956.

244. "The Copperhead." *Journal of Defense and Diplomacy*, October 1983, 49–52.

245. "The Queen's Gunners: British Field Artillery." *Artillery Trends*, August 1961, 8–13.

246. "The Soviet Union's Missile Artillery." *Artillery Trends*, December 1959, 24–27.

247. Thompson, W.F.K. "Yom Kippur and the Artillery." *Journal of the Royal Artillery*, September 1974, 81–89.

248. Toftoy, Holger N. "Army Missile Development." *Army Information Digest*, December 1956, 10–34.

249. Turbe, G. "Far-Reaching Changes in French Artillery." *International Defense Review*, February 1987, 403–4.

250. U.S. Army Command and General Staff College. *Division Artillery*. Fort Leavenworth, KS: U.S. Army Command and General Staff College, 1935.

251. ———. *Organization and Employment of a Division, Soviet (Russian) Army*. Fort Leavenworth, KS: U.S. Army Command and General Staff College Press, 1934.

252. ———. *Tactics and Techniques: Artillery*. Fort Leavenworth, KS: The General Service School Press, 1922.

253. ———. *Tactics and Techniques: Artillery*. 2 vols. Fort Leavenworth, KS: The General Service School Press, 1925.

254. U.S. Army Field Artillery School. *Artillery in Combat*. Fort Sill, OK: Field Artillery School, 1944–1945.

255. ———. *Development of Field Artillery Materiel*. Fort Sill, OK: Field Artillery School, 1934, 1939, and 1940.

256. ———. *Field Artillery Fundamentals*. Fort Sill, OK: Field Artillery School, 1942.
257. ———. *History of the Development of Field Artillery Materiel*. Fort Sill, OK: Field Artillery School, 1941.
258. ———. *Index to Publications Pertaining to Field Artillery Materiel*. Fort Sill, OK: Field Artillery School, 1941.
259. ———. *Materiel (Elementary)*. Fort Sill, OK: Field Artillery School, 1932, 1934, and 1937.
260. ———. *Tactical Employment of Field Artillery*. Fort Sill, OK: Field Artillery School, 1943.
261. U.S. Cavalry School. *Other Arms: Field Artillery with Cavalry*. Fort Riley, KS: QMC Plant, 1931.
262. U.S. Ordnance Department. *Handbook of Artillery*. Washington, DC: Government Printing Office, 1920, 1921, and 1931.
263. U.S.S.R. *Course in Artillery, Book One*. Translated by U.S. Army. Moscow: Military Publishing Office of the Ministry of the Armed Forces of the U.S.S.R., 1948.
264. ———. *Course in Artillery, Book Four*. Translated by U.S. Army. Moscow: Military Publishing Office of the Ministry of the Armed Forces of the U.S.S.R., 1947.
265. U.S.S.R. Ministry of Armed Force. *Course in Artillery*. Translated by U.S. Army Ordnance Intelligence Service. Aberdeen Proving Grounds: U.S. Army Ordnance Intelligence Service, 1955.
266. Van Creveld, Martin. *The Transformation of War*. New York: The Free Press, 1991.
267. Ventham, Philip, and David Fletcher. *Moving the Guns: The Mechanization of the Royal Artillery*. London: H.M.S.O., 1990.
268. Vigman, Fred K. "The Theoretical Evaluation of Artillery after World War I." *Field Artillery Journal*, January-February 1976, 21–23, 38. Reprinted from *Military Affairs*, Fall 1952.
269. Wakefield, Ken. *The Fighting Grasshoppers: U.S. Liaison Aircraft Operations in Europe, 1942–1945*. Leicester, England: Midland Counties Publications, 1990.
270. Walker, Paul F. "Precision Guided Weapons." *Scientific American*, August 1981, 37–45.
271. War Department (U.S.). *Design and Manufacture of Artillery Systems*. Translated by E. K. Larmen. Washington, DC: Intelligence Department, 1949.
272. ———. Military Intelligence Service. *The Organization and Tactical Employment of Soviet Ground Forces (July 1944)*. Washington, DC: War Department, 1946.
273. Watteville, H. G. de. "Russian Artillery: 1941–1945." *Journal of the Royal Artillery*, April 1947, 133–44.
274. Weathersby, Russell A. "The Field Artillery Group in Support of the Corps and Field Army, 1942–1953." Master's thesis, U.S. Army Command and General Staff College, 1965.
275. Weinert, Richard P., Jr. *A History of Army Aviation: 1950–1962*. Edited by Susan Canedy. Fort Monroe, VA: Office of the Command Historian, 1991.
276. Weyl, A. R. "Development of the V–2." *Military Review*, May 1949, 81–84.
277. ———. *Guided Missiles*. London: Temple Press, 1949.
278. Williams, P. D. "Space-Age Artillery." *Journal of the Royal Artillery*, March 1971, 4–11.

279. Willmott, H. P. *The Great Crusade: A New Complete History of the Second World War*. New York: The Free Press, 1990.

280. Wilson, Arthur R. *Field Artillery Manual*. Menasha, WI: George Banta Publishing Company, 1925–1926.

281. Wrixon, Tim. ''RPV Development Accelerates.'' *Jane's Defence Weekly*, 27 September 1986, 687–90.

282. Yant, Martin. *Desert Mirage: The True Story of the Gulf War*. Buffalo: Prometheus Books, 1991.

283. Zaloga, Steven J. ''The Mechanized Threat.'' *Field Artillery Journal*, January-February 1987, 39–42.

284. Zaloga, Steven J., and Arnold Meisner. *U.S. Mechanized Firepower*. London: Arms and Armour Press, 1987.

Part III

Biographical Profiles

This section contains brief biographical sketches of individuals who made significant contributions to the development of field artillery tactics, doctrine, organization, or materiel. Other people certainly advanced development of field artillery, but the following people stand apart from the rest.

Alexander, Edward P. (1835–1910). Confederate army officer during the American Civil War. Born in Washington, Georgia, Alexander graduated from the U.S. Military Academy in 1857, was commissioned in the engineers, and served on the faculty of the U.S. Military Academy for the next three years except for a brief time when he served in the Mormon Expedition to Utah in 1858. In June 1861 he accepted a captain's commission in the Confederate army and served on the staff of General P.G.T. Beauregard as an engineer and chief of signal service. One year later in November 1862, he took command of an artillery battalion in General James Longstreet's corps and was promoted to colonel after serving only a short period of time as a major.

During the American Civil War, he gained a reputation as a competent field artilleryman. At the Battle of Fredericksburg (December 1862), he commanded the batteries on Marye's Heights and swept the plain below of Union soldiers. Later at Chancellorsville (May 1863), he accompanied General Stonewall Jackson's famous flanking movement and positioned the batteries that helped shatter the Union right flank. At the Battle of Gettysburg (July 1863) as commander of the artillery reserve of General James Longstreet's corps, he had the duty of silencing Union batteries on the third day of the battle so that General George Pickett's units could attack. Lacking sufficient ammunition, Alexander's batteries could not provide effective support because they had to stop firing. As a result, the charge failed. Years after the war, he wrote *Military Memoirs of a Confederate* (1907), which provided his insights into Confederate field artillery practices.

Armstrong, Sir William G. (1810–1900). English inventor. Born in Newcastle upon Tyne, Northumberland, Armstrong invented high-pressure hydraulic machinery and revolutionized the design and manufacture of artillery. Armstrong abandoned his law practice in Newcastle in 1847 to devote all of his time to scientific experimentation. He founded an engineering works at Elswick-on-Tyne to build hydraulic cranes.

Armstrong soon became well-known for ordnance production by shrinking metal rings onto a steel barrel to produce hooped rifled artillery, revolutionizing artillery materiel. In 1854 Armstrong introduced a rifled breechloader. Eventually, Armstrong's firm became one of the leading arms manufacturers in Europe by buying out Sir Joseph Whitworth's (q.v) company in 1898 and Vickers' Sons and Maxim in 1909 to become W. G. Armstrong, Whitworth, and Company, Ltd. During World War I, Armstrong's company manufactured over 13,000 field artillery pieces.

Bacon, Roger (1214?–1294). English scholastic philosopher and scientist and a celebrated teacher. Born in Ilchester, Somersetshire, Bacon was educated at Oxford and the University of Paris. He remained in Paris after completing his studies and taught for a time at the University of Paris. Soon after his return to England in about 1251, he entered the religious order of the Franciscans and settled at Oxford, where he carried on active studies and did experimental research, mainly in alchemy, optics, and astronomy.

As early as 1261, he knew that the chemical mixture of saltpeter, sulfur, and powdered charcoal, eventually known as black powder, would explode. Even though Bacon saw that its fiery use could destroy a city, he apparently did not see a military purpose for black powder as a force for hurling projectiles.

Barry, William F. (1818–1879). U.S. Army officer. In 1838 he graduated from the U.S. Military Academy and was commissioned in the 2nd Field Artillery Regiment and later served in the Mexican War of 1846–1848. Because of his reputation as an artilleryman, he participated on the board that revised light artillery tactics for the U.S. Army in 1858–1859.

Subsequently, as chief of artillery of the Army of the Potomac under General George B. McClellan, he labored tirelessly along with Colonel Henry J. Hunt (q.v.) to improve the army's field artillery. When he began his work in July 1861, the army had only 30 guns, 400 horses, and 650 men. By March 1862 the artillery of the Army of the Potomac had 520 guns, 11,000 horses, and 12,500 trained artillerists. After the Peninsular Campaign of 1862, he asked to be reassigned and transferred to Washington, D.C., to serve as inspector of artillery for the armies of the United States and chief of artillery of the defenses of Washington. In March 1864 he was relieved and appointed chief of artillery under General Ulysses S. Grant. Later, he served as chief of artillery under General William T. Sherman during the Atlanta Campaign. After the Civil War he organized and headed the U.S. Army Artillery School at Fort Monroe, Virginia, from 1866 to 1877.

Benét, Stephen Vincent (1827–1895). U.S. Army officer. Born in St. Augustine, Florida, to a son of a prominent figure of Spanish descent, Benét entered a private school in Alexandria, Virginia, at the age of twelve. On the strength of his academic achievements there, he gained admittance to the University of Georgia but left shortly thereafter to enter the U.S. Military Academy in 1845. After graduating third in his class of 1849, Benét served in a variety of posts at the Watervliet, Frankford, and St. Louis arsenals and also taught at the U.S. Military Academy. By 1864 he had been made commandant of the Frankford Arsenal, a position that he held for five years. A year later, he was awarded brevets of major and lieutenant colonel for faithful and meritorious service as an ordnance officer during the Civil War.

In June 1874 the U.S. Army appointed Benét brigadier general and chief of ordnance. During his tenure as chief of ordnance, he oversaw critical improvements in field artillery. On the basis of successes in Europe with steel breechloaders, he decided that the army had to convert from wrought-iron muzzleloaders to steel breechloaders. As a result of his efforts, the U.S. Army adopted the M1885 3.2-inch steel breechloader in the mid–1880s. Even though the weapon lacked a recoil system, used black powder, and was, therefore, behind European technology of the 1880s, it represented a significant development in American field artillery.

Bessemer, Sir Henry (1813–1898). British inventor. Born in Charlton, Hertfordshire, England, Bessemer was largely self-educated. Although he was a productive inventor, his most famous contribution was his process for manufacturing steel. By blowing blasts of air over molten ore in a furnace that he invented in the 1850s, Bessemer not only refined the ore by removing the impurities to improve its quality but also raised the annual steel production in England at a fraction of the cost of the old method. The Bessemer process made steel field artillery a reality. Because of his contributions to science, he was knighted in 1879.

Bishop, Harry G. (1874–1934). U.S. Army field artillery officer. Born in Grand Rapids, Michigan, Bishop received an appointment to the U.S. Military Academy in 1893, from which he graduated in 1897. Upon graduation, he received a commission as a second lieutenant, was later promoted to first lieutenant in 1899 while he was in the 2nd Field Artillery Regiment, and was promoted to captain in 1901 in the Artillery Corps. When the field artillery was separated from coast artillery in 1907, he transferred to the field artillery, where he commanded a battery until 1910. In 1912 he was promoted to major in the 5th Field Artillery Regiment at Fort Sill, Oklahoma, and later commanded field artillery troops on the Mexican border in 1916. That year, he was promoted to lieutenant colonel, and was promoted to colonel in 1917.

During the 1920s and early 1930s, Bishop was one of the most progressive thinkers in the U.S. Army's field artillery. He strongly advocated motorization of divisional artillery, and wanted organic air observation for the field artillery, in order to take advantage of the long ranges and defilade positions of the new field artillery being fielded and to find more targets. As Chief of Field Artillery in the U.S. Army from 1930 to 1934, Major General Bishop made steady efforts to prepare the field artillery for motorization and organic air. However, motorization and organic air observation did not come until the eve of World War II. He wrote *Elements of Modern Field Artillery* (1914) and *Field Artillery: King of Battle* (1935), which was published posthumously.

Bomford, Henry (1780–1848). U.S. Army officer. Born in New York City, Bomford graduated from the U.S. Military Academy in 1805 and was commissioned a second lieutenant of engineers. From 1805 to 1808 he was assistant engineer on the fortifications in New York harbor. He was promoted to first lieutenant in 1806, to captain in 1808, to major in 1812, and to lieutenant colonel in 1815. He headed the Ordnance Department from 1821 to 1832, when he was promoted to colonel and placed in command for the newly organized, independent Ordnance Corps. From 1842 through 1848, he was also inspector of arsenals, ordnance, arms, and munitions of war for the U.S. Army.

During his career, he was the foremost expert on ordnance in the United States and worked actively to introduce a system of American artillery. He devised a cannon in the 1840s with the characteristics of a howitzer and mortar that he named Columbiad after the poem by Joel Barlow. Although he was best-known for his invention of this cannon for coastal defenses, which could fire solid shot and explosive shell and replaced seacoast howitzers in the 1840s, he supported mounting the U.S. Army's field batteries as horse artillery after years of neglect, and training the gun crews. Because of Secretary of War Joel Poinsett's and Bomford's efforts late in the 1830s, the U.S. Army's field artillery was modernized early in the 1840s and employed in the Mexican War of 1846–1848.

Braun, Wernher von (1912–1977). Rocket scientist. Born in Wirsitz, Germany, von Braun became fascinated with the possibilities of using rockets to explore outer space and in 1930 joined a group of amateur rocket enthusiasts in Berlin. He studied engineering at technological institutes in Zurich and Berlin and earned a degree from the latter in 1932. That same year, he became chief of a rocket research station established by the German army. In 1932 he was awarded a Ph.D. by the University of Berlin. In 1936 Adolf Hitler became interested in rockets as potential weapons and ordered the construction of a large research facility at Peenemünde, where von Braun and other scientists developed the Vergeltungswaffe Zwei (V–2) by 1938. Over the next six years, von Braun and his colleagues enlarged and improved the missile until it could carry a warhead of nearly one ton almost 200 miles in five minutes. Shortly before the war ended, von Braun and a large number of his colleagues fled before the advancing Soviet army and surrendered to U.S. forces.

In 1945 he came to the United States and became technical assistant director at the U.S. Army's proving ground for missiles at White Sands, New Mexico, and was later made director of the missile research facility at Huntsville, Alabama. Under his supervision, the Redstone, Jupiter-C, Juno, and Pershing rockets were developed. He pressed for a program of space exploration and wrote numerous papers, articles, and books on the subject and later became director of the George C. Marshall Space Flight Center at Huntsville, where he worked on the Saturn V employed in the Project Apollo program for a manned moon landing.

Brewer, Carlos (1890–1976). U.S. Army field artillery officer. Born in western Kentucky, he entered the U.S. Military Academy in 1909 and graduated in 1913. Upon graduation, Brewer joined the 3rd Field Artillery Regiment at Fort Sam Houston, Texas, and served along the Mexican border during the Mexican Revolution. In August 1916 he returned to the U.S. Military Academy, where he joined the Department of Mathematics. After attending the Field Artillery School at Fort Sill, Oklahoma, in 1926–1927 and the U.S. Army Command and General Staff College at Fort Leavenworth, Kansas, in 1927–1928, he returned to the Field Artillery School in 1928 as an instructor with the rank of major. From 1929 to 1933 he served as director of the Department of Gunnery.

As director of the department, Brewer made his most significant accomplishments. After reading about the exploits of Lieutenant Colonel Neil Fraser-Tytler with observed indirect fire during World War I, he revamped the U.S. Army field artillery's unobserved and observed fire early in the 1930s so that it was simpler than World War I techniques and so that it would be more responsive for mobile warfare. One big advantage of Brewer's method was that a few specialists could perform the necessary technical operations for the entire battalion so that field artillery fire could be massed rapidly and effectively. He retired in 1950 with the rank of major general after serving successfully as combat commander in the 6th Armored Division and commander of the 46th Field Artillery Group during World War II.

Coehorn, Menno van (1641–1704). Dutch military engineer and general. He was born in Leeuwarden, Friesland, and entered military service at the age of sixteen. At the siege of Grave in 1673, his engineering skills became evident when he harassed the French garrison with a small mortar that he invented and that was subsequently named the Coehorn mortar. He excelled in design and construction of fortifications and in siege techniques. His *New Fortifications* (1685) and other books were studied widely in Europe.

Congreve, Sir William (1772–1828). British artillerist. Born in Woolwich, England, Congreve was educated at Trinity College, Cambridge, where he studied law and graduated in 1793. He was the eldest son of Sir William Congreve, comptroller of the Woolwich Laboratory and inventor of mounted field artillery. Early in the 1800s, on the basis of the successful use of rockets in India for military purposes, he developed three calibers of rockets, a 1-pounder, a 3-pounder, and a 6-pounder. The 3-pounder had a range of 1,800 yards, and the 6-pounder had a range of 2,300 yards. However, the effective ranges of the rockets were about 600 to 700 yards. Because the warheads were incendiary, they would ignite whatever the rocket struck. To ensure that the rockets would adhere to whatever they hit, Congreve designed the warheads with sharp points.

Taking advantage of his father's position as comptroller of the Royal Laboratory at Woolwich, Congreve the younger was able to manufacture his rockets and later have them officially adopted by the British army. In the spring of 1805, his father, now a general, arranged with the Master General of Ordnance, John Pitt, second earl of Chatham, to initiate large-scale production of his son's rockets. Congreve's rockets set Boulogne ablaze in 1806 and Copenhagen afire in 1807 during the Napoleonic Wars. Impressed by the British successes with

Congreve's rockets, other European armies adopted similar ones, thus making Congreve's rocket the standard to imitate.

The younger Congreve also wrote *A Treatise on the General Principles, Powers, and Facility of Application of the Congreve Rocket System as Compared with Artillery* (1827) and *A Concise Account of the Origin and Progress of the Rocket* (1807). In the former, he described the use of his rocket in land and sea warfare.

Du Teil, Chevalier Jean (1738–1820). French artilleryman. In 1778 he published *De l'Usage de l'Artillerie Nouvelle dans la Guerre de Campagne* (On the Use of the New Artillery). He opened his book by declaring that the great improvements recently made in French artillery—the Gribeauval (q.v.) reforms—had revolutionized mobility and had made new tactics not only possible but also desirable to exploit the new technology. He suggested that the guns should open up the battle and then move rapidly to enfilade the enemy line with massed fire from huge batteries. In short, massing fire was the key to victory according to Du Teil.

During his career, Du Teil and his older brother, Baron Joseph Du Teil, had frequent contact with Napoleon. The younger Du Teil was an officer in the Regiment of La Fère into which Napoleon was commissioned in 1785. This regiment was later sent to the artillery school at Auxonne, which was commanded by Baron Du Teil. Later, after his family was exiled from Corsica, Napoleon found the younger Du Teil in command of the artillery of the Army of Italy. Du Teil made him an assistant. In this capacity Napoleon had his first opportunity to play an important part, by commanding the artillery at the siege of Toulon. Thus, Chevalier Du Teil had a direct influence upon Napoleonic field artillery tactics that were so successful during the first years of the nineteenth century.

Edward III (1312–1377). King of England. Born at Windsor, Edward was the eldest son of Edward II and Isabella of France, through whom he and subsequent kings of England laid claim to the throne of France. After his father's abdication late in 1326, Edward III was crowned king of England in January 1327 at the age of fourteen. In reality, however, his mother and Roger Mortimer, a previously exiled baron, ruled in his name for the next three years. After surprising, capturing, and executing Mortimer at Nottingham for treason in 1330, he forced his mother into retirement and then turned to the conquest of France.

In 1337 England declared war on France, with hostilities beginning two years later. After a brief, indecisive foray into the Low Countries, Edward proclaimed himself king of France. From 1340 until a disastrous campaign in 1359–1360, Edward's armies moved from victory to victory against the French. During the Battle of Crécy in 1346, he employed artillery on the battlefield. Although the artillery pieces did not contribute to the English victory, Edward III established an important precedent—the use of artillery on the battlefield—that subsequent commanders imitated. Later, in the siege of Calais in 1347 during his invasion of France, he proved that artillery was a valuable siege weapon.

Frederick the Great (1713–1786). King of Prussia. Frederick inherited a fine army of eighty thousand and a full war chest, as well as a centralized and subservient state. He quickly acquired proficiency in military arts and had an intellectual approach to war. Because Prussia was vulnerable to likely enemies, he depended upon an active, mobile army to blunt the assaults of an enemy before it could gather strength, and relied upon the offensive. In an age when generals still preferred the solid tempo of a siege to battle, Frederick was willing to take risks.

While he was king of Prussia, he dramatically revolutionized his country's artillery to help offset the numerical superiority of his enemies. He introduced cartridge ammunition, standardized his guns so that they served a particular function based upon size and weight, and in 1759 developed horse artillery in which the gun crew rode on the horses pulling the field gun to serve as a strategic reserve or to support the cavalry. His horse artillery and improved field artillery permitted the king to mass fire in an unprecedented manner, establishing himself as a leading artillerist of the mid-eighteenth century.

Goddard, Robert (1882–1945). American inventor and professor. Born in Worcester, Massachusetts, Goddard had a genteel upbringing, and in early childhood felt the excitement of the post–Civil War industrial revolution, when Worcester factories were producing machinery and goods for a rapidly expanding country. From childhood on, he displayed great curiosity about physical phenomena and had a bent toward inventiveness. He read in physics and mechanics and dreamed about great inventions.

In 1908 he began his long association with Clark University, Worcester, where he earned his doctorate, taught physics, and carried out rocket experiments. In his laboratory he proved that thrust and consequently propulsion could take place in a vacuum. He was also the first to develop a rocket motor employing liquid fuels, as used in the German V–2 rocket. In achieving lift-off with the small rocket engine that he built in his laboratory in the early 1920s, Goddard carried his experiments further than did Russian and German space pioneers of the day. He was the first to shoot a liquid-fuel rocket faster than sound. He obtained the first patent of a steering apparatus for the rocket and developed the first pumps suitable for rocket fuels, self-cooling rocket motors, and other components of an engine designed to carry a person into outer space. In 1945 he died at the beginning of the jet age.

Gribeauval, Jean-Baptiste (1715–1789). French artillerist. Possibly the greatest artillery reformer since Gustavus Adolphus (q.v.), Gribeauval entered the French army in 1732 and was promoted to officer rank in 1735. In the Seven Years' War (1756–1763) he was attached to the Austrian army and served as a general of artillery. With his experience of the Seven Years' War as a background, he started his work of rationalizing French artillery equipment in 1765. He discarded the multitude of calibers and patterns of guns and retained only 4-, 8-, and 12-pounders as field artillery, made to standard specifications and designed for certain functions. His reforms, which included training, paved the way for the successes of Napoleon and his marshals during the early years of the nineteenth century.

Guk, Karl (1846–1910). Russian army officer. He was educated in the Polotsk Cadet Corps and the Pavlovsk Military and Mikhailovsk Artillery academies. He served in the Brest Litovsk fortress artillery from 1866 to 1871. Here, he probably came into contact with the early indirect fire techniques for siege artillery. He then worked in the St. Petersburg ordnance workshop and the Kiev arsenal. In the Russo-Turkish War of 1877–1878 he commanded a siege battery and participated in the bombardment of Kars, where he employed primitive indirect fire methods. In 1881 he became deputy inspector of training at the Mikhailovsk Artillery Academy as a lieutenant colonel and was promoted to inspector and colonel in 1885. He was promoted to major general in 1895, made a member of the Artillery Committee in 1899, and was promoted to lieutenant general in 1903.

As a lieutenant colonel at the Mikhailovsk Artillery Academy, he wrote a book, *The Covered Fire of Field Artillery* (1882), that described the modern system of fire direction that was adopted by most armies. He outlined using a compass, aiming point, and forward observer to engage unseen targets and argued that rifled steel field artillery would be prime targets for enemy fire, that they had to be sited in defiladed positions to protect them, and that this required an effective method of indirect fire. He illustrated the use of aiming points, dealt with problems of crest clearance, and devised an elaborate system of signals for passing corrections from the observation post to the guns. He also predicted correctly that when the guns retired behind the masking hills, the forward observers would become prime targets. Guk's work was equal to that of the most advanced artillery thinkers in Europe and was an original contribution to artillery technique.

Gustavus Adolphus (1594–1632). King of Sweden. When Gustavus Adolphus came to the throne in 1611 at the age of sixteen, Sweden was in the middle of a war. Confronted with the weaknesses of his troops, he recognized that they had to be shaped into a sturdy and self-reliant fighting force by being properly organized, equipped, led, and trained. Like Maurice of Nassau, a great military reformer and thinker who radically reformed the army of the Netherlands, Gustavus Adolphus believed in starting from the basics and initiated a bold plan to reform the Swedish army. Although he continued to employ mercenaries, drawn from England or Scotland, he regularized the procedure of conscription that had evolved over the previous half century to ensure that it was fairer, that it provided a regular supply of men, and that the conscripts were of high quality. These reforms produced the most homogenous and close-knit infantry in Europe at the time. At the same time, Gustavus Adolphus demanded that the soldiers be paid regularly to dignify their status and to prevent looting in war and desertion in peacetime.

His most innovative reforms, however, focused on equipment. He decreased the number of pikemen and the weight of the musket to make it less burdensome in combat. Concurrently, he boldly reformed his artillery. Because he recognized that his artillery had to play a greater role on the battlefield, he introduced a light, mobile gun, divided his mobile artillery into field and siege, attached 4-pounders to the infantry for close support, trained his gun crews to perform intricate movements rapidly, and adopted cartridge ammunition for rapid loading and firing. At Breitenfeld in 1631, he demonstrated the value of trained gun crews and mobile artillery when he defeated Tilly. However, attitudes among other commanders, his death in 1632, and the state of technology prevented his reforms from becoming widespread in Europe for many years.

Hale, William (1797–1870). English ordnance expert. Hale was born in Colchester, Essex, England, in modest circumstances. His father was a baker, and his maternal grandfather was a versatile educator and writer on diverse subjects that ranged from comets to algebra. In the 1820s Hale showed a penchant for mechanics and took out a patent in 1827 for an internal-screw vessel. The ship utilized a crude form of jet propulsion, sucking water into the vessel with the steam-powdered screw and then discharging it in order to drive the ship forward.

Late in the 1830s, Hale moved near the Royal Arsenal at Woolwich. While here, he first became aware of Sir William Congreve's (q.v.) work with rockets and started work on his own. In the 1840s Hale introduced far-reaching improvements in rockets. He eliminated the cumbersome guidesticks and adopted stickless, spin-stabilized projectiles. Hale improved production methods by using a hydraulic press to drive the propellant into the case and employed steel for rocket cases. Ironically, major technological advances during the Industrial Revolution of the mid-nineteenth century—breechloading and rifling—improved field artillery's accuracy, rate of fire, and range so much that Hale's developments were eclipsed and rockets fell out of favor until World War II.

Hohenlohe-Ingelfingen, Prince Kraft zu (1827–1892). Prussian artillerist. Born in Upper Silesia, he was commissioned a second lieutenant in 1845 and attended the Artillery and Engineer School in 1845–1846 and the War College in 1851–1853. He was promoted to major in 1858, lieutenant colonel in 1861, and colonel in 1865. During the Austro-Prussian War of 1866, he served as commander of the reserve artillery of the Guard Corps. After the war, he was promoted to major general in 1868 and later served as commander of the Guard Corps Artillery Brigade. During the Franco-Prussian War of 1870–1871, he was inspector of artillery. He was promoted to lieutenant general in 1873 and served as commanding general of the 12th Division from 1873 to 1879.

He was a prolific military writer, and his *Letters on Artillery* (1890, translated by N. L. Walford) was particularly influential in Europe during the last three decades of the nineteenth century. An ardent supporter of indirect fire, he pointed out that the field artillery had to be sited in defilade positions to protect it from hostile artillery and small arms fire.

Hunt, Henry J. (1819–1888). U.S. Army field artillery officer. After graduating from the U.S. Military Academy in 1839, he was commissioned a second lieutenant in the 2nd Artillery Regiment, later served in the Mexican War of 1846–1848, and was promoted to captain in 1852 and major in 1861. Together with Captain William F. Barry (q.v.) and Captain William H. French, Hunt rewrote the U.S. Army's field artillery tactics to produce a drill manual entitled *Instructions for Field Artillery* (1860) that served as the basis of Civil War field artillery drill for Union and Confederate armies and was reprinted three times during the 1860s.

On the basis of his reputation as an excellent artillery officer, he was promoted to colonel in 1861 after the Civil War had begun, was made chief of the artillery reserve for the Army of the Potomac, and eventually was made a brigadier general in the U.S. Volunteers. During his tenure, he developed an artillery reserve of field guns that served effectively in massing fire and replacing pieces knocked out of action. His most notable achievements was helping to mass fire at Malvern Hill (July 1862) and Gettysburg (July 1863). In effect, his reserve was the forerunner of corps artillery in the twentieth century.

Knox, Henry (1750–1806). Major general and Chief of Artillery. Forced to quit the Boston Public Latin School because of economic necessity, Knox served an apprenticeship to Wharton and Bowles Booksellers. In 1771 he opened the London Bookstore in Boston, which became a popular meeting place for British officers. Siding early with the patriot cause, Knox reported military intelligence that he overheard in his bookstore to patriot leaders. Knox obtained a textbook knowledge of artillery and military engineering by reading from his large stock of military treatises and books. He joined a militia artillery company in Boston and became second in command when it was reorganized.

Because of his knowledge and prewar experience, on 17 November 1775 he was placed in command of the Continental Army's artillery, a position that he held during the rest of the American Revolution. During the winter of 1775–1776, he retrieved captured cannon and mortars from Fort Ticonderoga and dragged them cross-county to Boston, where they fortified the colonial positions surrounding the British and forced their evacuation. Under Knox's direction, the Continental Army organized its artillery into regiments, trained officers and men, and standardized its weapons to give it an artillery that rivaled that of the British.

Kosciusko, Thaddeus (1746–1817). Army officer. Born in Poland, he was educated at the Jesuit College of Breese and the Royal Military School in Warsaw, where he became first king's cadet. He graduated in 1769 as a captain and received a scholarship to study engineering and artillery at Mézières, France. Upon returning to Poland in 1774, he found little opportunity to apply his military skills and subsequently went back to France. In 1776 he borrowed money for a trip to America after being excited by the American Revolution. For his efforts in fortifying the Delaware River, Congress commissioned him as a colonel of engineers in October 1776 and sent him to Fort Ticonderoga, where he served in the Northern Army and devised an innovative plan to fortify the surrounding hills with artillery and interlocking fields of fire. Although General Horatio Gates approved the plan, he was replaced by General Philip Schuyler before work could begin. Schuyler canceled Kosciusko's plans and ordered him to strengthen the existing works. Later, Kosciusko reinforced the area around Saratoga with redoubts and entrenchments and placed the heaviest concentration of artillery on Bemis Heights, which covered all the avenues of approach. Next, he fortified West Point, where he set up a network of interlocking forts and batteries with fields of fire that could cover each other, the rivers, and the approach to the south of West Point. In his final assignment of the war, Kosciusko was posted to the Southern Army as chief engineer in August 1780.

Noted for serving as a volunteer officer in the Continental Army during the American Revolution, he was promoted to brigadier general and was given five hundred acres of land for his services. He also completed one of the first field artillery drill manuals for the U.S. Army. At the request of the United States Military Philosophical Society, Kosciusko returned to the United States after living in France for several years and wrote *Maneuvers of Horse Artillery* (1808), which became the basis of American horse artillery drill in the U.S. Army until the 1820s.

Krupp, Alfred (1812–1887). Industrialist and arms manufacturer. Born in Essen, Germany, he succeeded his father, Friedrich, who had founded a small iron firm in 1810. Beginning with only seven workers after his father's death, Alfred built the firm into a giant by making use of the newest steel-making techniques, by tough discipline, and by dedicated leadership. Much of Krupp's success was based on the reputation that he gained when he invented a seamless steel railway wheel that won wide adoption.

Even so, Krupp earned much of his acclaim by producing steel cannon. In 1847 he cast his first steel cannon, a small 3-pounder, but he still continued to specialize in producing fine steel suitable for dies, tools, and machine building. A flawless, two-ton steel block displayed at the Crystal Palace of the Great Exhibition in 1851 demonstrated Krupp's ability. To boost steel production to meet Germany's needs, Krupp adopted the Bessemer converter and later the Siemens-Martin open hearth.

Large-scale armament production began after Krupp received the first Prussian arms order in 1858. Krupp steel cannon with their horizontal breechblocks contributed to Prussian victories over Austria in 1866 and France in 1871 and made him known as the cannon king.

Liechtenstein, Prince Joseph Wenzel (1696–1772). Austrian artillerist. Born in Prague in 1696 and educated at the Carolinum, he became a knight of the golden fleece in 1740. Impressed by the eighty-two well-served Prussian cannon that presented a striking contrast to antiquated Austrian guns at the Battle of Chotusitz (1742), Liechtenstein developed a comprehensive plan for reforming Austrian artillery. On the basis of this, Maria Theresa made him director general of the whole artillery arm in 1744. However, war intervened before Liechtenstein could begin work. After the war he established an artillery school for practical and theoretical training near Budweis, and exercise and experimental grounds to the north at Moldauthein. Here, he built laboratories and oversaw experiments in ballistics, cartridges, rapid fire, mines, and mortar bombs. He also standardized the mobile artillery at 3-, 6-, and 12-pounders, complemented them with light field howitzers in 1753, expanded and militarized the personnel, created artillery fusiliers, and published artillery drill manuals. His reforms led to the achievements of Austrian artillery during the initial battles of the Seven Years' War and caused Jean-Baptiste Gribeauval (q.v.), who was in the service of the Austrian army, to take note and eventually reform French artillery to keep abreast of its Austrian counterpart.

Maximilian I (1459–1519). Holy Roman emperor and Hapsburg archduke of Austria. Born in Weiner Neustadt, he was the eldest son of the emperor Frederick III. He married Mary of Burgundy, whose father, Charles the Bold, duke of Burgundy, died and left Mary sole heir of Burgundy, France-Comté, and the Low Countries. During much of his time as emperor, he was engaged in warfare trying to regain Mary's territories, which were seized by France, and to preserve the Hapsburg empire.

This almost constant warfare led Maximilian I to rationalize his artillery in the 1490s in order to make it more effective in combat. To eliminate the confusion caused by a multiplicity of calibers, he reduced the number of calibers and classified his artillery as siege or field, with the heavier pieces serving as siege artillery. In addition, Maximilian ordered the general use of iron shot, promoted gunnery training, organized his artillery into a separate branch of his army, and established a precedent that other Europeans imitated.

Moore, Dan T. (1877–1941). U.S. Army field artillery officer. Moore entered the U.S. Army in 1899 as a second lieutenant of infantry and served in the Spanish-American War of 1898 as a second lieutenant of Connecticut Infantry. In 1901 he transferred to the artillery, where he acquired an excellent reputation. During the years 1900–1908, a group of U.S. Army field artillery officers recognized that several foreign countries were far ahead of American field artillery tactics and techniques. With this in mind, the War Department sent the able and energetic Captain Dan T. Moore of the Sixth Artillery Regiment abroad to study in foreign artillery schools. He spent 1908–1909 visiting artillery schools in England, Holland, and Austria, and was a student at the German artillery school in Jüterbog.

After Moore returned from Europe, the U.S. Army dispatched him to Fort Sill, Oklahoma, to make the arrangements for establishing the School of Fire. As the first commandant of the school, Moore drew up the regulations and training programs to train American field artillery personnel in the use of indirect fire, which was just being introduced in European armies. As a field artilleryman, Captain Moore was acknowledged throughout the branch for his technical knowledge, progressive ideas, and energy in modernizing the field artillery. After World War I Moore resigned as a colonel and received a commission of brigadier general in the Officers' Reserve Corps.

Mordecai, Alfred (1804–1887). U.S. Army ordnance officer. Mordecai received his early schooling at home and at the local academy founded and directed by his father. After graduating from the U.S. Military Academy in 1823 as first in his class, he was commissioned a second lieutenant of engineers. He served in this capacity for nine years as an instructor at the U.S. Military Academy, as an assistant engineer in the building of Fortress Monroe and Fort Calhoun at Hampton Roads, and as an assistant to the chief of engineers. As a result of an army reorganization, Mordecai was appointed a captain of ordnance.

Although he held many important positions during his army career, such as command of the arsenals at Frankford, Pennsylvania, Washington, D.C., and Watervliet, New York, he was best-known for his work on the Ordnance Board on which he served from 1839 to the eve of the American Civil War (1861–1865). Organized to standardize and systematize army ordnance equipment, the board expanded its mission to include testing, evaluating, and designing virtually all the army's equipment.

As a part of this impressive effort, Mordecai along with fellow board members formulated the first complete system of artillery for the U.S. Army, including field artillery. Mordecai played a vital role in collecting, analyzing, and organizing the data studied by the board and wrote the board's final report, which was published under his name in 1849 as *Artillery for the United States Land Service*. His observations on the Crimean War in the 1850s directly led to the adoption of the Napoleon M1857 12-pounder gun-howitzer. He also did most of the work on the first *Ordnance Manual* (1841) ever published by the U.S. Army, was solely responsible for its second edition in 1850, and contributed to several editions of the published ordnance regulations. He eventually rose to the rank of colonel.

Muller, John (1699–1784). German-born mathematician. Although he was born in Germany and received his early training there, he moved to England later in his life. He wrote his first book in 1736, apparently while he was working at the Tower of London. Five years later, he was appointed headmaster of the Royal Military Academy at Woolwich. When the school was reorganized and the cadet company formed, he was appointed professor of fortifications and artillery and held that post until he retired in 1766.

As a mathematician, he believed that artillery should be based upon mathematical principles. In 1759 his *Treatise on Artillery* appeared while he was teaching at the Royal Military Academy. The book was persuasive, and it influenced British artillery. In addition, the *Treatise* affected American artillery designs during the American Revolution. A pirated edition was edited by John Norman and printed in Philadelphia in 1779 and was dedicated to George Washington and Henry Knox (q.v.). Most likely, Muller's treatise was the only manual available during the American Revolution. As a result, the Americans followed its prescriptions closely.

Napoleon III, Charles Louis Napoleon Bonaparte (1808–1873). King of France. Born in Paris to the third son of Louis Bonaparte, he was brought up at Geneva, Augsburg, and his mother's residence, the Swiss castle of Arenenberg on Lake Constance. On the death of the duke of Reichstadt, the only son of Napoleon I in 1832, he became the head of the Napoleonic dynasty. After a stormy political career he dissolved the constitution with the help of the military in 1851 and assumed the imperial title the following year.

Although Napoleon III was more noted for his political activities, he made a significant contribution to the development of French field artillery. In the middle of the nineteenth century he had developed a 12-pounder gun-howitzer with a trajectory that fit between that of a gun and that of a howitzer and that combined the firepower of heavy French field ordnance with the mobility of lighter guns. This Napoleon gun-howitzer replaced French 24- and 32-pounder howitzers and 8- and 12-pounder guns to give the French a powerful mobile field gun. The Americans later adopted the gun as the M1857 Napolean 12-pounder howitzer and used it extensively in the American Civil War.

Parrott, Robert P. (1804–1877). American inventor. In 1824 Parrott graduated from the U.S. Military Academy but resigned his commission in the U.S. Army to become superintendent of the West Point Foundry, Cold Spring, New York, in 1836. He leased the foundry in 1839 and operated it until 1867.

In the early nineteenth century rifled artillery was impractical because the metals available would not withstand the high firing pressures. In 1861 Parrott produced the first American rifled field gun by combining two metals. Using casting techniques of shrinking a wrought-iron cylinder over a cast-iron breech, he invented a practical way of strengthening a field artillery piece without increasing its weight. Parrott guns were widely used during the American Civil War.

Robins, Benjamin (1707–1751). Mathematician and military engineer. Born in Bath to a poor Quaker, Robins demonstrated mathematical ability at an early age and later in life applied himself to modern languages. As a teacher of mathematics, he devoted his time to tutoring individuals and not classroom instruction. For some years, he taught pure and applied mathematics and physical science but found it to be confining. As a result, he gave up teaching and became an engineer. He constructed mills, bridges, and harbors and worked to make rivers navigable.

In the meantime, Robins began studying gunnery and fortification. In 1742 his *New Principles of Gunnery* appeared, which explained the dynamics of exterior and interior ballistics and advocated rifled artillery and small arms. This work was the result of many experiments that he had conducted on the resisting power of air to swift- and slow-moving objects. He also invented a ballistic pendulum, a device for measuring the velocity of a projectile, and was considered the father of modern gunnery.

Rodman, Thomas J. (1815–1871). American inventor. Rodman graduated from the U.S. Military Academy in 1841. As an American army officer, he made one of the most significant inventions in the history of artillery propellants. During the late 1850s and early 1860s, he developed prismatic gunpowder by shaping grains of powder so that each grain could burn simultaneously from the inside and the outside. This kind of powder burned evenly and provided for controlled expansion of the gases in a gun rather than the sudden explosion that was typical of corned gunpowder.

Rodman also invented a system of casting cannon around a hollow core through which a stream of water was run to cool the metal. Thus, the bore surface cooled first, resulting in concentric layers of metal cooling and shrinking one upon another. This created a stronger cannon than either casting it as a solid block of metal and then boring out, or using molds to form a hollow cylinder. During the American Civil War, he commanded the arsenal at Watertown, Massachusetts, and was brevetted to brigadier general in 1856. At the end of the war, the U.S. Army placed him in command of the arsenal at Rock Island, Illinois.

Saint Barbara (ca. 200 to 300 A.D.). According to legend, Saint Barbara, the beautiful daughter of Dioscorus, a nobleman of the Roman Empire, lived in Nicomedia in Asia Minor in the third or fourth century A.D. To limit Barbara's exposure to Christianity and encourage her development as a zealous pagan, her father shut her up in a tower that was lighted by only two windows. From these windows she looked out upon the surrounding countryside and marveled at the living things. She concluded that they had to be part of a master plan and that the idols of wood and stone that her parents worshipped were false. After obtaining instruction in Christianity, she was baptized. In token of her faith, while her father was away she made a third window to symbolize the Holy Trinity. On his return Dioscorus asked why Barbara had made this change, and Barbara acknowledged her conversion to Christianity. He delivered her to a Roman magistrate, who also failed to persuade her to abandon Christianity. Dioscorus then took his daughter to a high mountain, where he beheaded her. As he descended the mountain afterward, he was caught in a sudden, violent storm. In a blinding flash he was consumed by lightning.

As a logical consequence, Barbara came to be regarded as the sainted patroness of those in danger from thunderstorms, fire, explosions, or any sudden death. Given the questionable reliability—misfires, muzzle bursts, and explosions—of cannon, early cannoneers sought the protection of Saint Barbara.

Schwarz, Berthold (fl. 1300s). Legendary inventor of gunpowder. Rather than spending his time on his clerical duties, Schwarz explored the mystical properties of alchemy in the search of a way of making gold from cheaper metals. He mixed sulfur, saltpeter (potassium nitrate), quicksilver (mercury), oil, and lead. Every time that he mixed this combination he blew off the lids of his kettles. Around 1320 he realized that this mixture that could blow lids off kettles could be used as an explosive to propel projectiles through the air.

Shrapnel, Henry (1761–1842). British army officer. Born in Bradford-on-Avon, he later received a commission in the Royal Artillery in 1779 and served in Newfoundland, Gibraltar, the West Indies, and Flanders. He devoted his talent for invention to improvements in ordnance. Around 1784 he began working with cannonballs filled with lead musket balls and gunpowder, which was called spherical case shot and later shrapnel. Shrapnel designed his projectile as an antipersonnel round to complement canister. The fuse detonated the round to burst the casing, but the lead balls continued forward in a cone-shaped pattern. His invention was extremely lethal against standing infantry in the open and was therefore soon adopted by most European armies to replace canister as the preferred antipersonnel round. Although he was made a lieutenant general in 1837, he never won any special recognition for his invention, which was widely used from the beginning of the nineteenth century through World War I.

Snow, William J. (1868–1947). U.S. Army field artilleryman. Born in Brooklyn, New York, he entered West Point at the age of seventeen and graduated in 1890. Upon graduation he was commissioned in the artillery and during the first years of his service was stationed near his home at Forts Hamilton, Wadsworth, and Slocum. In 1898 he graduated from the Artillery School at Fort Monroe, Virginia, and for the next thirty years was a leading pioneer in the development and improvement of the field artillery. In 1906 he became secretary of the Mounted Service School, Fort Riley, Kansas, and held that position for one year until he was transferred to the 6th Field Artillery, where he was the adjutant. After founding the Field Artillery Association in 1910, he became the first editor of the *Field Artillery Journal* in 1911.

His greatest accomplishments came when he was Chief of Field Artillery from

1918 through 1927. In this capacity during World War I, he oversaw the rapid expansion of the field artillery and invigorated the arm by organizing training schools. He promulgated a comprehensive training program that accelerated the flow of trained field artillerymen to units in the American Expeditionary Force, and created replacement centers. Equally important, he stressed observed indirect fire as opposed to unobserved indirect fire, the latter of which dominated field artillery tactics during the war. He retired from the U.S. Army in 1927 as a major general.

Tartaglia, Niccolo (1500?–1557). Italian mathematician and founder of ballistics and the art of gunnery. In 1537 he published the first scientific treatise on gunnery. Following this, he invented the gunner's quadrant, which aided aiming by determining the tube's elevation.

Torstensson, Lennart (1603–1651). Swedish artillerist. At the age of fifteen Torstensson became a page to Gustavus Adolphus (q.v.) and accompanied him on the king's campaigns of 1621–1623. Possibly because Torstensson showed some promise, Gustavus sent him to study under the great Maurice of Nassau, who was one of the first commanders to develop the full potential of artillery, especially in siege warfare, by standardizing calibers and training officers in its specialized use. After returning to Sweden, Torstensson served for three years in the Prussian campaigns, where he demonstrated his artillery skills, and became commander of the First Artillery Regiment in the Swedish army at the age of twenty-seven.

Although Gustavus Adolphus's reforms were not complete, Torstensson assumed command of the most modern artillery of the time. Following Maurice's example, Gustavus Adolphus reduced his artillery to 3-, 12-, and 24-pounders, designated them for specific roles, and stressed mobility. At Breitenfeld in September 1631, Torstensson showed what well-trained artillery could do. Increased mobility allowed the 3-pounders to accompany the infantry when it counterattacked and helped overwhelm Tilly's army. Because of his direction of the artillery, Gustavus Adolphus promoted Torstensson to general in 1632. At the battle of the Lech in April 1632, Swedish batteries commanded by Torstensson provided a massive barrage that forced the enemy to take cover and permitted the Swedish army to cross the River Lech. Eventually, Torstensson was promoted to field marshal in 1641, even though his health was poor. His genius as an artilleryman and the rise of Sweden as a major power made Torstensson one of the leading artillerymen in history.

Vallière, Jean-Florent de (1667–1759). French artilleryman. As inspector general of the French artillery, Vallière adopted a new artillery system that was named after him. The materiel of Louis XIV was powerful but heavy and immobile. During the 1730s, Vallière systematized French artillery, classifying cannon by the weight of the shot and assigning missions accordingly. At the same time he reduced the number of different calibers in French service to simplify logistical support, giving the French the only unified artillery system in Europe. For example, 4- and 8-pounders were classified as field artillery and larger cannon as siege or garrison. His artillery was sturdy but heavy and difficult to maneuver on the battlefield and, therefore, provided little improvement over Louis XIV's. Vallière's son retained the system when he became inspector general of the French artillery.

As long as the infantry was immobile, the Vallière system served well. With the rise of more mobile infantry and artillery in the mid-eighteenth century, younger artillery officers clamored to follow Austrian and Prussian developments that had produced cannon with the ability to maneuver with mobile infantry. When Gribeauval (q.v) became inspector general of French artillery in the 1770s, he replaced the Vallière system but not without strong opposition.

Vauban, Sébastien de Prestre de (1633–1707). French military engineer and theorist. In the third war of the Fronde of 1651–1653, Vauban threw in his lot with the Frondeurs by accepting a cadetship from their military leader, Louis II of Bourbon. In 1653 he was taken captive because he had already shown great aptitude in building field fortifications. Two years later, he became the king's engineer and started serving in a corps specializing in fortification and siegecraft.

After the War of Devolution in 1667–1668, in which he played a major role in the siege and capture of Douai, Tournai, and Lille, he received the opportunity to test his theories on fortification and siegecraft when he assisted in reforming the French army. Under his direction the French army built a chain of fortifications that were equipped as bases for an army on the march and that could not be defeated except by time-consuming siege. He was directly responsible for building thirty-three new fortresses and remodeling many others. He created the fortified naval bases of Brest, Dunkirk, Le Havre, Rochefort, and Toulon in response to Colbert's plans to create a strong French navy.

In the Dutch War of 1672–1678, Vauban first used his system of parallels and zigzag trenches for dealing with sophisticated fortresses. His technique played a significant role in defeating bulwarks at Philippsburg in 1688, Mons in 1691, Namur in 1692, and Charleroi in 1693. Because of the success of Vauban's method of siegecraft, other armies in Europe soon adopted it.

Wadsworth, Decius (1768–1821). First Chief of Ordnance in the U.S. Army. Displaying an intellectual bent early in life, Wadsworth completed his primary education rapidly and matriculated in Yale College at the age of fourteen. He graduated in 1785. Although he considered teaching, he returned to Yale and earned a master's degree in 1788. Later he became a lawyer but soon left the legal field. In searching for an alternative career, he looked to the U.S. Army, where Major General Anthony Wayne was attracting attention with his preparations for a major campaign on the frontier in the 1790s.

When Congress created the Corps of Artillerists and Engineers in 1794, Wadsworth successfully sought a commission in the new organization and was appointed a captain in 1794. Assigned to the artillery when a separate Corps of Engineers was established, Wadsworth persuaded Secretary of War Henry Dearborn to approve his transfer to the engineers.

Although Wadsworth performed well in the Corps of Engineers, his most noteworthy achievements occurred when he was Commissary General of Ordnance from 1812 to 1815 and Chief of Ordnance from 1815 to 1821. Appointed colonel and Commissary General of Ordnance, Wadsworth developed a small but highly efficient corps. He created arsenals in Albany, New York, and Pittsburgh, Pennsylvania, and undertook efforts to standardize the U.S. Army's artillery in order to facilitate logistical support in the field and to end the practice, continued since the Revolution, of depending upon a motley collection of artillery. Following the War of 1812, he advocated adopting a stock-trail field carriage based upon British ones captured during the War of 1812, and even designed one. He presented his carriage to Secretary of War John C. Calhoun in 1817. Before accepting Wadsworth's recommendation, however, Calhoun directed an ordnance board to study its potential. Impressed with the Gribeauval carriage, the board rejected the stock-trail carriage for the more famous French one, although the British carriage made it obsolete. Even so, the War Department adopted Wadsworth's principle of standardizing the artillery in 1816, but over two decades passed before the U.S. Army had a system of artillery, because of disputes over proper gun metal and carriages.

Ward, Orlando (1891–1972). U.S. Army field artillery officer. Born in Macon, Missouri, Ward received a commission as a second lieutenant of cavalry upon graduation from the U.S. Military Academy in 1914. Following graduation, he served along the Mexican border and participated in the Pershing punitive expedition into Mexico in 1916. After seeing duty during World War I with the 3rd Division staff and graduating from the Army General Staff College, Langres, France, he returned to the United States, where he served as professor of military science and tactics at the University of Wisconsin from 1919 to 1923. During the 1920s, Ward served in various army assignments, was promoted to major in 1920, and graduated from the U.S. Army Command and General Staff College in 1926 as the distinguished graduate.

His most notable contribution to the field artillery came when he received orders to the Field Artillery School, Fort Sill, Oklahoma, to become the director of the Gunnery Department in 1931. Here, he continued the work initiated by Major Carlos Brewer (q.v.) on improving fire direction for the U.S. Army's field artillery. Unlike Brewer, Ward found a way to centralize command and control of a battalion's three firing batteries to make them one firing unit. This breakthrough paved the way for the field artillery to shift massed fires around the battlefields of World War II with ease and effectiveness. This work, along with excellent service during World War II, led to a promotion to major general in 1942.

Wheeler, Charles B. (1865–1946). U.S. Army ordnance officer. After graduating from the U.S. Military Academy in 1887, he was assigned to the artillery for the first three years of his career. After being promoted to first lieutenant in 1890, he transferred to ordnance. For the next six years, he served at Sandy Hook Proving Ground and Watertown Arsenal, New York. He eventually rose to brigadier general in 1917 after serving successfully in the lower ranks but retired in his permanent grade of colonel in 1919.

During duty in the Office of Chief of Ordnance from 1896 to 1906, he played an instrumental role in developing the U.S. Army's M1902 3-inch field gun, the answer to the French M1897 75mm field gun. Mounted on a long recoil carriage, the M1902 rivaled its French counterpart. The M1902 had a interrupted-screw breech and had a range of 6,250 yards. Because of the desire for commonality when the United States entered World War I in 1917, the U.S. Army abandoned the M1902 in favor of the French M1897. As a result, none of the 640 M1902s in existence at the outbreak of the war ever saw combat action. Wheeler also designed the M1904 4.7-inch gun, which could throw a 60-pound projectile up to 7,600 yards. In his work with the M1902, the M1904, and other field artillery material at the beginning of the twentieth century, Wheeler played a critical part in introducing modern, rapid-fire field guns to the U.S. Army.

Whitworth, Sir Joseph (1803–1887). English mechanical engineer. He started as a mechanic in Manchester and London, discovered the method of making a truly plane surface by scraping, and established himself as a toolmaker in Manchester. Between 1840 and 1850 he developed his standard system of uniform measures and gauges and designed measuring machines that were accurate up to one millionth of an inch. After experimentation with rifles, he perfected a hexagonal-barreled rifle exceeding in range and accuracy the Enfield, which was in use at the time. However, in 1869 the War Office refuse to adopt it.

In 1859 he invented a wrought-iron hoop field gun with a hexagonal bore that fired a hexagonal projectile, and became one of the first to develop a workable rifled breechloader. His rifled field gun threw a 250-pound projectile up to six miles. However, the Ordnance Board rejected it in 1865 because of the tendency of breechloaders to leak gases and to explode upon firing. Nevertheless, his method of compression of ductile cast steel for ordnance was a great advance and came into general use.

Ziska, John (1376–1424). Hussite general of the Bohemian civil wars of the fifteenth century. Born to a minor country nobleman, he was brought up at the court of Vaclav IV and lost an eye at an early age. The Hussites were a puritan sect following the reformer John Huss, who had acquired great influence in Bohemia at the end of the fourteenth century. Their popular appeal was based upon their religious reforms as well as nationalism, which was on the rise among the Czechs.

Ziska gained fame for his tactics and his field victories as a commander when he was totally blind. During a temporary armistice in the struggles between the Hussites and Catholics, Ziska visited a Hussite fortress, where he was responsible for developing a formidable military community. While he was there, he noticed the effective way that the Russians and Lithuanians employed wagon barricades as a defense against heavy cavalry. He projected the idea much further. The Hussites traveled in large horse-drawn wagons with armored sides and loop holes. When he had selected his ground, Ziska would circle his wagons, position light artillery pieces, which were carried in wagons on the march, in the gaps, and surround the wagons and artillery with a defensive ditch. This combination proved to be almost invincible. If the enemy did not show any inclination to attack, Ziska would send out his cavalry to harry the surrounding country until he could no longer be ignored. Generally, the wagon fortress firepower broke the enemy after a few abortive charges. Thereupon, Ziska would counterattack with his pikemen and cavalry to drive the enemy from the field. Ziska's novel tactics led to many battlefield successes for the Hussites.

The following individuals were less prominent in the development of field artillery, and there is little biographical information available on them.

Chrystie, J. Walter (fl. 1920–1940). American industrialist. Chrystie developed self-propelled artillery during the 1920s.

Culmann, General Frederick (fl. 1910–1940). French army officer. During the 1930s, Culmann was considered the greatest living French authority on field artillery tactics and doctrine.

De Bange, Colonel C. Ragon (fl. 1870–1890). French army officer. De Bange was known for his interrupted-screw breech system, which many European field artillerymen of the 1870s and 1880s thought was better than Krupp's (q.v.) horizontal sliding wedge block.

Herr, General Frederick Georges (fl. 1910–1930). French army officer. Herr was a prominent field artilleryman during World War I known for his contributions to doctrine.

Langlois, General Hippolyte (1839–1912). French army officer. Langlois played a significant role in developing field artillery doctrine during the 1890s.

Rimaihlo, Colonel Émile (fl. 1890–1920). French army officer. Rimaihlo played a critical part in developing the French M1897 75mm field gun during the 1890s.

Rohne, General Heinrich (1842–1937). German army officer. Rohne served as editor of *Artilleristische Monatshefte* (Artillery Monthly) and was a leading advocate of high-explosive shell.

Rouquerol, General Jean Gabriel Marie (fl. 1900–1940). French army officer. Rouquerol was a leading advocate of indirect fire during the 1890s. He insisted that direct fire had been made obsolete by the development of rapid-fire field guns.

Toftoy, Major General Holger N. (fl. 1950s). American army officer. Toftoy was a major leader in the development of American field artillery rockets and missiles in the 1950s.

Voronov, Marshal Nikolai N. (1899–1968). Soviet army officer. Voronov played a major role in developing Soviet artillery into an effective combat arm in World War II and was a leader in the development of artillery and missiles in the postwar years.

Wahrendorff, Baron Martin (fl. 1840–1850). Swedish industrialist. Wahrendorff developed a system of rifled field artillery during the 1850s and developed a projectile of soft lead that fit tightly in the bore of the weapon to engage the rifling.

Appendix 1
Chronology of Important Dates

1346: Artillery was employed on the battlefield for the first time when the English used several small, carriageless artillery pieces against the French during the Battle of Crécy.

1453: The siege of Constantinople by Mohammed II breached the heavy city walls and led to the downfall of the Byzantine Empire.

1494–1495: Charles VIII's train of mobile field and siege artillery quickly and effectively breached Italian fortifications, forcing them to be remodeled or new ones to be built designed to withstand gunpowder artillery.

1537: Niccolò Tartaglia published *Nova Scientia*, laying the foundations of scientific gunnery.

1620s: King Gustavus Adolphus of Sweden introduced his leather 4-pounder field gun in an effort to have mobile artillery on the battlefield.

1671: Louis XIV of France founded one of the first permanent artillery regiments in Europe.

1742: Benjamin Robins published *New Principles of Gunnery*, establishing the science of ballistics.

1759: Frederick the Great of Prussia introduced horse artillery, which was soon adopted by most European armies.

1762: Jean-Baptiste Gribeauval began his artillery reforms in the French army that revolutionized field artillery and set the standard for other European armies to follow.

1778: Jean Du Teil published his influential *On the Use of the New Artillery*, in which he advocated massing field artillery in unsurpassed proportions because of the light, mobile field artillery introduced by Gribeauval.

1784: Lieutenant Henry Shrapnel of the British army introduced spherical case shot.

1804: Sir William Congreve began his work with rockets.

1846: Major Giovanni Cavalli of Piedmont, Italy, introduced one of the first workable breech-loading rifled field guns.

1873: Alfred Krupp developed one of the first workable recoil systems for field artillery.

1884: Paul Vieille of France introduced a smokeless propellant that was suitable for military use and soon adopted by the field artillery. It increased range because it was more powerful than black powder, and gave off less smoke than black powder did.

1897: The French M1897 75mm field gun appeared and revolutionized field artillery.

1904–1905: The successful employment of indirect fire in the Russo-Japanese War encouraged European and American field artillerymen to adopt it to protect their guns from counterbattery and small arms fire.

1932–1934: The U.S. Army Field Artillery School developed the fire direction center, revolutionizing fire direction in field artillery.

1953: The Americans tested Atomic Annie, a 280mm field gun that fired an atomic projectile and made atomic artillery a reality.

1961: The first field artillery computer to go into service was the U.S. Army's Field Artillery Digital Automatic Computer.

1973: The successful use of remotely piloted vehicles by the Israeli army during the Arab-Israeli War for battlefield artillery target acquisition in the Bekaa Valley spurred on the Europeans and the Americans to develop their own remotely piloted vehicles for artillery observation.

1978: The initial production of the Copperhead precision-guided munition began.

1991: The Persian Gulf War vindicated the massive rearmament programs of the 1980s that saw the infusion of high technology into field artillery.

Appendix 2
List of Great Bombards

Name	Date (Century)	Metal	Weight in tons	Length (inches)	Weight of shot (lbs)
Dulle Griete of Ghent	15th	Wrought iron	13	197	700
Mons Meg	15th	Wrought iron	5.7	159	330
Cannon of Mohammed II	15th	Bronze	18.7	204.7	1,250
Malik-i-Mydan, Great Gun of Beejapore	16th	Bronze	40	170.6	1,000
Tzar Pooscka, Great Gun of Moscow	16th	Bronze	38.5	213	2,240

Source: O.F.G. Hogg, *Artillery: Its Origin, Heyday, and Decline* (Hamden, CT: Archon Books, 1970), 210, 212, 219, and 285.

Appendix 3
English Ordnance, 1640s

Name	Caliber (inches)	Weight (pounds)	Length (feet)	Weight of shot (pounds)
Cannon Royal	8	8,000	8	63
Cannon	7	7,000	10	47
Demi-cannon	6	6,000	12	27
Culverin	5	4,000	11	15
Demi-culverin	4.5	3,600	10	9
Saker	3.5	2,500	9.5	5.25
Minion	3	1,500	8	4
Falcon	2.75	700	6	2.25
Falconet	2	210	4	1.25
Robinet	1.25	120	3	.75

Source: O.F.G. Hogg, *Artillery: Its Origin, Heyday, and Decline* (Hamden, CT: Archon Books, 1970), 272.

Appendix 4
French Ordnance, 1666

Name	Weight (pounds)	Weight of Charge (pounds)	Weight of Projectile (pounds)
Whole Cannon	6,100	22.5	33.33
Great Culverin	3,400	10	15
Quarter Cannon	2,800	8.5	12
Saker	1,700	4.25	6
Sixteenth Cannon	1,100	2.5	3
Thirty-second Cannon	750	1.5	1.5

Source: O.F.G. Hogg, *Artillery: Its Origin, Heyday, and Decline* (Hamden, CT: Archon Books, 1970), 283.

Appendix 5
Select American Field Artillery Pieces, 1860s

Caliber	Projectile	Range in Yards	Weight of Shot	Smoothbore of Same Caliber
M1841 6-pounder, Smoothbore	shot	1,525	6lbs.	N.A.
M1857 Napoleon, Smoothbore	shot	1,680	12lbs.	N.A.
	sph case	1,135	12lbs.	N.A.
	shell	1,300	12lbs.	N.A.
10-pounder Parrott, Rifled	shell	2,000	9.75lbs.	3-pounder
	shell	5,000	9.75lbs.	3-pounder
20-pounder Parrott, Rifled	shell	2,100	18.75lbs.	6-pounder
	shell	4,400	18.75lbs.	6-pounder

Source: Albert Manucy, *Artillery through the Ages: Emphasizing the Types Used in America* (Washington, DC: Government Printing Office, 1949), 56; Boyd L. Dastrup, *King of Battle: A Branch History of the U.S. Army's Field Artillery* (Fort Monroe, VA: Office of the Command Historian, 1992), 144.

Appendix 6
Select Field Artillery, 1897–1945

Name	Caliber in Inches	Weight in Pounds	Weight of Shell in Inches	Range in Yards
British 13-pounder	3.0	2,236	13.0	5,900
French 75mm, M1897	2.95	2,510	16.0	8,500
British 4.5-inch Howitzer	4.5	4,030	35.0	7,000
British 18-pounder	3.3	3,507	18.0	11,100
British 25-pounder	3.45	3,968	25.0	13,400
Japanese 75mm	2.95	2,487	14.0	8,865
French 155mm Howitzer, M1917	6.1	7,275	96.0	12,575
French 155mm GPF	6.1	23,700	98.0	20,560
French 105mm Howitzer, M35	4.13	3,586	35.4	13,120
U.S. 105mm Howitzer, M2A1	4.13	4,980	33.0	12,500
Russian 152mm Howitzer, M43	5.98	7,938	88.0	13,550
Russian 122mm Gun, M31	4.80	15,652	55.1	22,825

Source: Ian V. Hogg, *Artillery in Color: 1920–1963* (New York: Arco Publishing, Inc., 1980), 181–82.

Appendix 7
Select Field Artillery, 1945–1991

Name	Caliber in MM	Traverse on Carriage	Range in Meters	Mount
UK Abbot	105	360 degrees	17,300	SP
USA M52	105	120 "	11,200	SP
USA M108	105	360 "	11,270	SP
USA M101A1	105	45 "	11,160	Towed
USSR M1955	122	60 "	21,000	Towed
China Type 54	122	45 "	11,800	SP
USSR M1974	122	360 "	15,300	SP
USSR M1946	130	50 "	27,000	Towed
USSR M1955	152	60 "	18,000	Towed
French F3	155	50 "	21,600	SP
French GCT	155	360 "	23,300	SP
Israel L/33	155	60 "	21,525	SP
Israel M72	155	360 "	20,500	SP
UK AS-90	155	360 "	24,700	SP
USA M109A2	155	360 "	18,100	SP
USA M109A6	155	360 "	45,000	SP
S. Africa G6	155	80 "	30,000	SP
Sweden FH77A	155	60 "	22,000	Towed
USSR M1975	203	45? "	30,000	SP

Source: Ian V. Hogg, *Artillery 2000* (London: Arms and Armour Press, 1990), 151–54; U.S. Field Artillery School, Program and Project Summary Sheets, 5 October 1992, 4-1–4-2.

Author Index

Subject Index

About the Author

BOYD L. DASTRUP, Command Historian, U.S. Army Field Artillery Center at Fort Sill, Oklahoma, is well-known for his expert knowledge about U.S. Army history. His books include *Crusade in Nuremberg: U.S. Military Occupation, 1945–1949* (Greenwood Press, 1985), and *U.S. Army Command and General Staff College: A Centennial History* (1981).